W9-CRV-316

THEORETICAL ASPECTS OF HOMOGENEOUS CATALYSIS

Catalysis by Metal Complexes

VOLUME 18

The titles published in this series are listed at the end of this volume.

THEORETICAL ASPECTS OF HOMOGENEOUS CATALYSIS

Applications of *Ab Initio* Molecular Orbital Theory

Edited by

PIET W.N.M. VAN LEEUWEN
Van 't Hoff Research Institute, University of Amsterdam, The Netherlands

and

KEIJI MOROKUMA
C.L. Emerson Center, Emory University, Atlanta, U.S.A.

and

JOOP H. VAN LENTHE
Theoretical Chemistry Group, University of Utrecht, The Netherlands

KLUWER ACADEMIC PUBLISHERS
DORDRECHT / BOSTON / LONDON

Library of Congress Cataloging-in-Publication Data

Theoretical aspects of homogeneous catalysis : applications of ab
initio molecular orbital theory / edited by Piet W.N.M. van Leeuwen
and Keiji Morokuma and Joop H. van Lenthe.
p. cm. -- (Catalysis by metal complexes ; v. 18)
Includes index.
ISBN 0-7923-3107-9 (acid-free)
1. Catalysis. I. Leeuwen, P. W. N. M. van (Piet W. N. M.)
II. Morokuma, K. (Keiji), 1934- . III. Lenthe, Joop H. van.
IV. Series.
QD505.T474 1995
547'.050595--dc20 94-31171

ISBN 0-7923-3107-9

Published by Kluwer Academic Publishers,
P.O. Box 17, 3300 AA Dordrecht, The Netherlands.

Kluwer Academic Publishers incorporates
the publishing programmes of
Martinus Nijhoff, Dr W. Junk, D. Reidel and MTP Press.

Sold and distributed in the U.S.A. and Canada
by Kluwer Academic Publishers,
101 Philip Drive, Norwell, MA 02061, U.S.A.

In all other countries, sold and distributed
by Kluwer Academic Publishers Group,
P.O. Box 322, 3300 AH Dordrecht, The Netherlands.

Printed on acid-free paper

Typeset in the United Kingdom
Printed in The Netherlands

TABLE OF CONTENTS

P.W.N.M. VAN LEEUWEN

PREFACE

Homogeneous catalysis plays an important role both in the laboratory and in the industry. Successful applications in industry involve new polymerisation processes with complexes of zirconium and related metals, new carbonylation processes employing palladium and rhodium, ring opening polymerisations, and new enantioselective isomerisation catalysts as in the preparation of menthol. Also in the synthesis of organic compounds in the laboratory highly selective homogeneous catalysts represent an irreplaceable part of the toolbox of the synthetic chemist. Examples of such reactions are cross-coupling (Ni, Pd), nucleophilic substitution of allylpalladium complexes, Heck reactions (Pd), asymmetric epoxidation, Wacker type reactions (Pd), asymmetric hydrogenations (Rh, Ru), reactions of chromium complexes, enantioselective reactions with Lewis acids, reactions with the McMurry reagent, etc. There is hardly any multistep organic synthesis that does not involve one of these metal catalysed reactions. Most of these catalysts have been developed by empiricism.

The metal catalysed processes consist of a series of elementary steps which often have been studied in isolation in organometallic chemistry. The knowledge of such elementary steps – effect of ligands, anions, coordination number, valence states – has greatly contributed to the development of improved catalysts for the reactions mentioned above. In addition to the empirical approach theoretical methods have given support and guidance to the development of improved processes. Often the key steps of a cycle escape from a direct observation and then theoretical contributions are even more wanted. Thus, we have seen in recent years a growth of both molecular mechanics and quantumchemical studies in homogeneous catalysis. Until a decade ago the quantumchemical studies mostly concerned extended Hückel methods which gave a clear qualitative insight in the symmetry of the orbitals involved in the elementary steps of the organometallic catalysts. Especially for those instances where empirical methods failed the need for more quantitative data on the elementary steps grew in order to guide the experimental work. In the last decade *ab initio* MO methods and Density Functional Theory methods have been successfully introduced for the study of elementary steps of cata-

P.W.N.M. van Leeuwen et al. (eds), Theoretical aspects of homogeneous catalysis, 1–2.

lytic cycles. The aim of this book is to familiarise the people who work on the development of homogeneous catalysts with the recent advances in this field. It is not the aim of this work to consider the methods in detail. The prominent elementary steps of homogeneous catalysis are dealt with and the reader will learn about the most up-to-date treatment of these steps.

We will see that quantitative predictions can be made for a variety of elementary steps. For a certain reaction a trend for ligand effects can be predicted as well. Most authors claim, under the best circumstances, an accuracy of the calculated activation energies of 3 kcal per mol. In terms of predicting "rates and selectivities of catalysts" this is still rather disappointing; both rates and selectivities may be off by two orders of magnitude. One would like to know selectivity enhancements of an even smaller magnitude. Furthermore, in a calculation we can consider only steps that are very similar. In catalysis often competitive reactions occur for which the pathways are not suited for a quantitative comparison, e.g. ionic species, solvent molecules or different substrates may be involved. More importantly, the actual catalysts contain far more atoms than can be handled by our computers in spite of the enormous progress made in the last two decades. The introductory Chapter 1 outlines some of the possibilities and limitations.

The study of oxidative addition reactions (Chapter 2) shows the importance of ligands and the geometry of the complexes formed in this reaction as compared to the gaseous metal atom. Chapters 3 and 4 deal with migratory insertion reactions which have received a great deal of attention in literature. In this instance the experimental knowledge on e.g. the effects of ligands on the rate of this reaction is much less developed although this reaction is of great industrial importance. This is clearly an example of a reaction for which a direct experimental study is difficult when a very fast catalyst is involved. Organometallic 2+2 reactions (Chapter 5) are important in metathesis and polymerisation reactions. Nucleophilic attack of coordinated alkenes is discussed in Chapter 6 (Wacker reactions). In the last chapter (7, epoxidation with manganese porphyrin catalysts) the reader is confronted with the occurrence of paramagnetic states which may well be typical of epoxidation/porphyrin reactions. Thus, a broad spectrum of organometallic catalytic reactions or their elementary steps is discussed employing also a broad spectrum *ab initio* methods thus giving an impression of the state of affairs. It is hoped that the experimental chemists will find useful concepts in this volume that may help in the development of better catalysts.

Piet W.N.M. van Leeuwen

P.H.M. BUDZELAAR AND J.H. VAN LENTHE

AN INTRODUCTION TO QUANTUMCHEMICAL ORGANOMETALLIC CHEMISTRY

1. Introduction

During the last two decades, computational organic chemistry has earned its place as a discipline complementary to experimental chemistry. Starting out as a somewhat esoteric occupation of sometimes doubtful relevance, it is now recognised as a valuable research tool used by many researchers in both industry and academia. The calculations have become accurate enough to be relevant for experimental chemists and they have become quick enough not to be a full-time occupation.

The continuing expansion of computer technology and associated improvement of computer software has made this possible. The improvements are nicely illustrated by a table from the GAUSSIAN brochure, showing the timing for Gaussian92 test job number 178[1]:

TABLE I
RHF/6–31G** Single point energy for tri-amino-trinitro-benzene (300 basis functions)

Program	Computer system	Approx. CPU time
Polyatom (c. 1967)	CDC 1604	200 years*
Gaussian 80	VAX 11/780	1 week*
Gaussian 88	Cray Y-MP	1 hour
	IBM RS/6000 Model 550	4.5 hours
Gaussian 92	Cray Y-MP	9 minutes
	Cray C90	4.5 minutes
	IBM RS/6000 Model 550	1 hour
	486 DX2/50	20 hours

* Ignoring memory and disk limitations.

In the area of organometallic chemistry, it is taking computational chemistry much longer to achieve similar importance. There are a number of reasons for this, which we will discuss in more detail below. Nevertheless even in this field "computer chemistry" is growing in importance. It has passed the stage of qualitative orbital diagrams and is now making at least semi-quantitative predictions that can help experimental chemists.

P.W.N.M. van Leeuwen et al. (eds), Theoretical aspects of homogeneous catalysis, 3–13.

The purpose of this book is to highlight some of the insights obtained from theoretical studies and make them available to researchers in organometallic chemistry and homogeneous catalysis. Therefore, attention in this volume is focused on results and interpretation, rather than on methodological aspects. Each of the following chapters focuses on a particular reaction type. Subjects are Oxidative Addition, Alkene Migratory Insertions and C-C Bond Formation, 2+2 reactions, Wacker-type reactions and Epoxidation.

The treatment is not uniform; different authors treat their subjects at different levels of sophistication, which is brought about by a combination of the requirements of the systems studied and hardware and software limitations, as well as by the personal preference of the authors. The chapters contain a significant amount of theoretical background themselves.

In this introductory chapter we summarise the history of computational chemistry in organic and organometallic chemistry. We consider briefly the theoretical methods useful in the study of organometallic compounds and the choices one has to make when using them. Finally, we discuss the kind of results one may obtain from a computational study and list some books, that may be of use to the reader, who is interested in the quantumchemical methods.

2. A history of theoretical organic chemistry

The earliest example of a "theoretical model" used in organic chemistry is probably the development by Le Bel and Van 't Hoff of the "tetrahedral carbon" model [2, 3]. This model, while purely descriptive, could be used to explain or at least bring some order into a large part of organic chemistry. When quantum mechanics was postulated in the beginning of this century, it was immediately applied to the covalent bonds of organic compounds, leading to qualitative understanding of the nature of the chemical bond [4].

The calculation of π-orbital energies using the Hückel method [5] in the 1930's is probably the first example of a "computational chemistry" approach. Soon thereafter the treatment was generalised to include σ-bonding, leading to the Extended Hückel method [6–9] in the 1950's. These qualitative methods provided a framework for discussing the electronic structure and chemical behaviour of organic molecules. They produced concepts like "forbidden" and "allowed" reactions, "frontier orbitals" and the famous Woodward-Hoffmann rules [10].

The advent of electronic computers provided the opportunity for *quantitative* calculations. Three different classes of methods, each with their own area of application, made their entry into organic chemistry, approximately simultaneously in the period of 1970–1980:

- *Ab initio* methods attempt to solve the Schrödinger equation in a fairly rigorous way. They are used to study the electronic structure in "small" molecules. Such calculations nearly always employ model systems, i.e. simplified versions of the molecules that are studied experimentally, to investigate some effects in isolation.
- *Semi-empirical* methods use most of the formalisms of *ab initio* methods, but replace the parts of the energy expression, that are difficult or time-consuming to calculate, by approximations fitted to give the best results for a set of reference molecules. These methods can handle larger systems than the *ab initio* approach and are used for "medium sized" molecules. They may fail for molecules with unusual bonding characteristics, which were not present in the "reference set", but they can be at least as accurate as the *ab initio* methods for standard molecules.
- *Force-field* calculations treat molecules as "ball and spring" classical systems, ignoring electronic structure and quantum mechanics altogether. Again parameters (bond strengths, steric repulsion, etc.) are fitted to reproduce experimental data. Since organic chemistry is dominated by localised, covalent bonds, force field methods can be very accurate in this area (better than either *ab initio* or *semi-empirical* methods), provided that the molecule studied is very similar to the ones used in the reference set from which the parameters were determined. Because of the very simple nature of the energy expressions *force-field* calculations can handle very large systems (up to 10^6 atoms). A serious disadvantage is that the user has to assign atom and bond types *a priori*, based on chemical knowledge or intuition, which degrades the predictive value of the theory.

All three approaches owe much of their success and acceptance to the development by dedicated research groups of widely distributed general purpose computer codes. For *ab initio* programs the premier example is the GAUSSIAN series of programs by Pople *et al.* [11, 12] and many others are now available like GAMESS, CADPAC, MOLPRO, TURBOMOL, SPARTAN, etc. For *Semi-empirical* programs the scene was set by the MINDO [13–15]-MNDO [16]-AM [17] series of programs by Dewar and others, and for *Force-field* calculations we may mention Allinger's MM [18–20] programs. Although there is a wealth of programs to choose from, considerable standardisation has been obtained in the field of computational organic chemistry.

3. A history of theoretical organometallic chemistry

Coordination and organometallic chemistry are much younger subjects than organic chemistry. The first organometallic compound was prepared around 1850 [21] and the first structurally characterised organo-transition metal compound was ferrocene, prepared in 1951 [22, 23]. Soon after its discovery, the Dewar-Chatt-Duncanson model was put forward [24, 25] to explain the nature of the interaction between the metal and its "ligands' in ferrocene and related compounds. From early in the seventies Hoffmann and others used the extended Hückel method to understand bonding and reactivity in organometallic compounds. As in organic chemistry, these calculations provided a framework for the classification of the interactions, as is exemplified by the famous "isolobal analogy" [26].

Following these early applications of computational methods to organometallic chemistry, one could have expected, in analogy to organic chemistry, a fairly rapid development of standardised *ab initio*, *semi-empirical* and *force-field* calculations. The primary reason this did not happen is probably the wide range of structures and bonding types observed in organometallic and coordination chemistry. A single metal atom can have a bewildering variety of oxidation states, coordination geometries and can participate in a choice of bond types, as is exemplified by the series of iron compounds $Fe(CO)_4^{2-}$, $Fe(CO)_5$, $[CpFe(CO)_2]_2$ (*Fe-Fe*), $[CpFe(CO)]_4$ (4 *Fe-Fe*), Cp_2Fe, $Fe(H_2O)_6^{2+}$, $(C_3H_5)Fe(CO)_3I$, Fe_2Cl_6, FeO_4^{2-}. It appears to be very difficult to devise a parametrization scheme for a *semi-empirical* or *force-field* method, that gives reliable results for each of these environments. Also, if possible at all, such a parametrization requires a large set of reference compounds and a corresponding large amount of thermochemical data, which are scarce for organometallic compounds. There are now a few "reasonable" semi-empirical methods for transition metals (e.g. ZINDO [27] and SINDO [28]), but their reliability can not bear a comparison with that of typical "organic" methods like MNDO. For *force-field* methods one might have to use a different "iron type" for each of examples mentioned above, leading to an inordinate number of force-field parameters to be determined. In practice, one either develops a dedicated set of parameters for a small group of very similar compounds, or alternatively one lets a program guess parameters "on the fly" without too much worry about accuracy. Clearly neither approach will result in a good general purpose computational method.

Ab initio methods do not suffer from the problems mentioned above and should thus be ideally suited for calculations on organometallic complexes.

While this may be true in principle, there do exist a number of problems that have contributed towards delaying the widespread use of *ab initio* methods in this area.

A first problem is that even the simplest model of an organometallic complex is often large compared to a model organic compound, if only because of the large number of electrons and valence shells of the metal atom. For example, ferrocene (96 electrons) cannot be simplified much, if we aim to understand its electronic structure; compare this to typical organic model molecules like ethene (16 electrons) or benzene (42 electrons) bearing in mind that *ab initio* calculations scale at least with N^3, the third power of the system size.

A second problem is that electron correlation may be quite important in organometallic chemistry. In view of the often small energy differences involved in homogeneous catalysis and the probably subtle differences in reaction mechanisms, sophisticated and hence expensive calculation methods are called for. Also, in particular for third-row transition metals relativistic effects can be significant. These can be handled at various levels of accuracy at the expense of even more computer time.

Thus it is understandable that the use of *ab initio* methods in organometallic chemistry has lagged behind their application in organic chemistry. Developments in computer algorithms and hardware open up new possibilities every year, so routine application of these methods is coming within reach now.

4. Methods and choices in computational organometallic chemistry

In this book mostly *ab initio* methods are used. We define this concept here pragmatically as methods, that do not require experimental information, as opposed to *semi-empirical* and *force-field* approaches, which require parameters that are fitted to reproduce experimental data. The value of the result of a calculation critically depends on the appropriateness of the quantumchemical approach chosen and there is still a large amount of research going on. There is no such thing as a generally applicable quantumchemical technique.

The customary way to view *ab initio* methods is as attempts to find approximate solutions to the time-independent Schrödinger equation [29]:

$$\mathbf{H}\Psi = E\Psi \tag{1}$$

Where **H** is the Hamiltonian, defining all relevant details of the chemical system (position and kind of the nuclei, number of electrons, etc.) and Ψ and E are the wavefunction and energy of the system, respectively. The nuclear motion is generally left out of the equation. This is the Born-Oppenheimer [30] or "clamped nuclei" approximation. It is such a common approximation that it is often not even mentioned.

The Hartree-Fock [31–35] or SCF method is the simplest such *ab initio* technique. A single set of occupied orbitals is combined into a determinant to define the wave function. An effective one-electron operator, the Fock operator, is employed to obtain unique optimal molecular orbitals. An attractive feature of the method is that it offers a readily interpretable orbital picture of the electronic structure of a molecule. Unfortunately, this simple approach cannot describe some bond breaking and forming processes, occurring in chemical reactions, correctly. Neither can it yield highly accurate results. The former specific shortcomings can be rectified by using Multi Configuration SCF methods like Generalized Valence Bond (GVB [36]) or more general MCSCF [37] approaches. Also at a cost of forfeiting the fact that the wave function adheres to the proper spin, one could use the Unrestricted Hartree-Fock method [38]. These methods still support a useful orbital picture of chemical bonding and reactivity, though the interpretation is somewhat more difficult than in the simple Hartree-Fock model. These orbital methods may be collectively referred to as HF models.

For quantitative results, it is usually necessary to go beyond the Hartree-Fock approximation and include correlation using treatments that may be collectively labelled as **HF+**. The more sophisticated methods are not only more expensive, in terms of computer resources, they also lack the simple orbital picture of the HF methods. Therefore the *understanding* may come from a HF calculation whereas the higher accuracy is achieved using some correlation treatment without trying to interpret the improvement too closely. There is an abundance of HF+ methods, ranging from simple perturbation theory (e.g. MP2 [39]) to sophisticated correlation treatments. This is still very much an area of active research, making it difficult to give general rules as to which methods perform best for which systems.

An approach that falls a bit outside this classification is Density Functional Theory (DFT) [40, 41]. In contrast to the *ab initio* methods discussed above, no attempt is made to solve the Schrödinger equation, but the energy is sought as a (formally unknown but existing) functional of the one-electron density. It does include some correlation and employs no fitted parameters, though there is an extending range of functionals to choose from. A problem is that formally, there is no way to describe proper spin states or excited electronic states, which may make application to open-shell transition metal complexes difficult.

A problem, which falls in a different category, is whether relativity is taken into account [42], either at great cost by explicit inclusion in the Hamiltonian or by applying relativistic corrections. Relativistic corrections tend to make atoms more compact, increase bond strengths and reaction barriers. For third-row transition metals relativistic corrections are nearly always required. For second-row metals they are needed if more subtle effects are considered as in comparing different spin states or if accurate inner-shell ionisation energies are sought, but they are not required for a description of most chemical reactions.

Apart from the formalism to be used for the calculation, there are a number of choices to make in performing a proper calculation. The most important one is the choice of basis set, which determines the flexibility electron density has to adapt to changing molecular situations. In contrast to the situation in organic chemistry no general agreement exists on which basis set is needed in a given situation. The basis sets, that are used in practice are in order of increasing flexibility, accuracy and cost:

- Minimal (e.g. STO3G): use only for qualitative calculations.
- Split-Valence (SV) or Double Zeta (DZ), e.g. 3–21G, 6–31G: gives reasonable results at the HF level. In order to obtain reliable estimates of bond angles, polarisation functions are required.
- Many special-purpose extended basis sets have been developed: these are needed to accurately calculate energies or sensitive properties like polarisabilities, NMR parameters, etc.

The number of basis functions that are included may be reduced by treating only the 'valence" electrons explicitly and replacing the inner shells of transition metals by effective core potentials (ECP's), assuming that these are not important for the chemistry. This is particularly attractive for third-row transition metals, since the ECP's may be adapted to include some relativistic corrections. ECP's like any approximations must be handled with care.

Obviously one has to balance the choices of method and basis sets. It does not make sense to employ a very sophisticated correlation treatment combined with a minimal basis set. Also the cost of a calculation must be taken into account. The amount of computer resources (time, memory, disk) that is required, rises as the third power of the number of basis functions (N^3) for the simpler *ab initio* approaches. For the more sophisticated methods the dependence of the computational expenditure on the system size may be like N^6 to N^7.

There is no agreement in the literature to what is the required combination in a given situation as may become clear from reading the other chapters in this book. There are few standards as yet and computational organometallic chemistry is still a developing and fascinating discipline.

5. Interpreting the results of *ab initio* calculations

A straightforward HF (or DFT) calculation produces an energy and a wave function, which may be used to interpret the bonding and to calculate a wealth of molecular properties. Often only the energy is actually used. This energy as a function of all or some of the atomic positions is a Potential Energy Surface (PES). Due to the generally large number of nuclear coordinates (degrees of freedom) it is expensive to sample a large part. Also interpretation of a PES is difficult, particularly if the number of degrees of freedom is large. One may derive molecular structure or a reaction path from it.

The **molecular structure** can be compared directly with experiment, for instance with a X-ray diffraction structure. In organic compounds one should expect agreements within a few hundredths of an Ångstrøm for bond lengths or within a few degrees for angles at the HF/split-valence level. For organometallic compounds, where bonding is often much "softer" and correlation corrections are more important larger deviations (up to ~ 0.2 Å and 10° for "soft bonds") are common.

Calculating a **reaction path** and especially the **transition state** (TS) gives the activation energy of a chemical reaction. Usually the system size precludes the calculation of absolute reaction rates, which ideally should be done quantum mechanically. Activation energies calculated at the HF level are usually too high, typically by some 10 kcal/mole for "allowed" reactions. In organic chemistry, reactions that are "forbidden" by spin or space symmetry usually do not take place at all. In organometallic chemistry ground and excited states are often so close, that formally forbidden reactions are actually quite easy, even though it is often not possible to describe them at the HF level. Therefore one should always try to follow the orbitals in going from reactants to products and be alert to possible orbital crossings that might require a GVB or MCSCF treatment. Also one should be aware of spin-orbit interactions, which, especially in transition metal compounds, may make "spin-forbidden" reactions quite possible.

Calculating a reaction path or part of a Potential Energy surface can provide the basic information for quantitative data on reaction rates, but it does not explain by itself why a particular reaction is easy or difficult. To understand energy differences or reaction rates attempts are made to analyze and compare the wave functions of reactants, products and transition states. HF wavefunctions can be analyzed in the same way as the Extended Hückel MO (EHMO) wavefunctions that Hoffmann and others used for their interpretation of organometallic chemistry [26]. In particular, frontier orbital arguments

(HOMO-LUMO interactions) are valuable for understanding attractive and repulsive interactions [43]. Interpretation of wavefunctions that are beyond the HF approximation, is more difficult.

The molecular properties, like charge distributions, dipole moments, equipotential surfaces or polarisabilities, which may be obtained from a HF or HF+ calculation can be very helpful in understanding the course of a reaction. Most interpretations in organometallic chemistry have centered on MO arguments, but there is little doubt that charge control can be at least as important as orbital control, particularly in the early stages of a reaction. Interpretations and predictions based on molecular properties, e.g. Molecular Electrostatic Potential Surfaces (MEPS) [44], have become quite important in pharmacochemistry [45] and there is no reason to believe, that they could not rise to the same status in organometallic chemistry.

6. Background reading

The following list gives a short description of some books discussing quantumchemical methods.

W. J. Hehre, L. Radom, P. v. R. Schleyer and J. A. Pople: *ab initio* Molecular Orbital Theory [46].
This standard guide gives a simple account of molecular orbital theory and describes various techniques used in practical calculations. It also give some insight in the way a computation and a quantumchemistry program is set up. Most of the information is pertinent to the GAUSSIAN series of programs, one of the most popular program packages.

P. W. Atkins: QUANTA, A Handbook of Concepts (second edition) [47].
This book provides a qualitative pictorial picture of concepts in quantum theory. Each subject is treated in generally less than a page as separate entry.

A. Hinchcliffe: Computational Quantum Chemistry [48].
A little booklet (only 110 pages) discussing in a practical way many of the subjects hinted at in this introduction.

A. Szabo and N. S. Ostlund: Modern Quantum Chemistry, Introduction to Advanced Electronic Structure Theory [49].
A graduate textbook focusing on the more formal and mathematical background of Hartree-Fock and sophisticated methods. Various methods are illustrated by calculations on very simple molecules.

R. McWeeny: Methods of Molecular Quantum Mechanics (second edition) [50].
A standard textbook at graduate to advanced graduate level for someone who really wants to get into Quantum Chemistry.

P. H. M. Budzelaar
Shell Research bv Badhuisweg 3
1031 CM Amsterdam, The Netherlands

J. H. van Lenthe
Theoretical Chemistry Group, Debye Institute
Utrecht University, Padualaan 14, 3584 CH Utrecht, The Netherlands

References

1. Gaussian Inc: Gaussian92 brochure (1993)
2. J.A. Le Bel, Bull. Soc. Chim. Fr. **22**, 337 (1874)
3. J.H. van 't Hoff: La Chimie dans l'Espace (1875)
4. L. Pauling: The nature of the chemical bond. Ithaca (1940)
5. E. Hückel, Z. Physik **60**, 423 (1930)
6. R. Hoffmann, J. Chem. Phys. **39**, 1397 (1963)
7. C. Sandorfy, Can. J. Chem. **33**, 1337 (1955)
8. K. Fukui, H. Kato and T. Yonezawa, Bull. Chem. Soc. Japan **33**, 1197 (1960)
9. K. Fukui, H. Kato and T. Yonezawa, Bull. Chem. Soc. Japan **33**, 1201 (1960)
10. R.B. Woodward and R. Hoffmann, J. Am. Chem. Soc. **87**, 395 (1965)
11. W.J. Hehre, W.A. Lathan, M.D. Newton, R. Ditchfield and J.A. Pople, Gaussian 70, Bloomington, Indiana (1970)
12. J.S. Binkley, R.A. Whiteside, R. Krishnan, R. Seeger, H.B. Schlegel, D.J. Defrees and J.A. Pople, Gaussian 80 (1980)
13. N.C. Baird and M.J.S. Dewar, J. Chem. Phys. **50**, 1261 (1969)
14. M.J.S. Dewar and E. Haselbach, J. Am. Chem. Soc. **92**, 590 (1970)
15. R.C. Bingam, M.J.S. Dewar and D.H. Lo, J. Am. Chem. Soc. **97**, 1285 (1975)
16. M.J.S. Dewar and W. Thiel, J. Am. Chem. Soc. **99**, 4899 (1977)
17. M.J.S. Dewar, E.G. Zoebisch, E.F. Healy and J.J.P. Stewart, J. Am. Chem. Soc. **107**, 3902 (1985)
18. N.L. Allinger, Adv. Phys. Org. Chem. **13**, 1 (1976)
19. N.L. Allinger, J. Am. Chem. Soc. **99**, 8127 (1977)
20. N.L. Allinger, Y.H. Yuh and J.-H. Lii, J. Am. Chem. Soc. **111**, 8551 (1989)
21. E. Frankland, J. Chem. Soc. 263 (1848)
22. T.J. Kealy and P.L. Pauson, Nature **168**, 1039 (1951)
23. S.A. Miller, J.A. Tebboth and J.F. Tremaine, J. Chem. Soc. 632 (1952)
24. M.J.S. Dewar, Bull. Soc. Chim. Fr. **18**, C79 (1951)
25. J. Chatt and L.A. Duncanson, J. Chem. Soc. 2939 (1953)
26. R. Hoffmann, Angew. Chem. **94**, 725 (1982)
27. W.P. Anderson, T.R. Cundarai, R.S. Drago and M.C. Zerner, Inorg. Chem. **1**, 29 (1990)
28. J. Li, P.C. de Mello and K. Jug, J. Comp. Chem. **13**, 85 (1992)
29. E. Schrödinger, Ann. Physik **79**, 361 (1926)
30. M. Born and J.R. Oppenheimer, Ann. Phys. **84**, 457 (1927)
31. D.R. Hartree, Proc. Camb. phil. Soc. math. phys. Sci **24**, 328 (1928)
32. V. Fock, Z. Phys. **61**, 126 (1930)

33. J.C. Slater, Phys. Rev. **35**, 210 (1930)
34. C.C.J. Roothaan, Rev. Mod. Phys. **23**, 69 (1951)
35. G.G. Hall, Proc. Roy. Soc. (London) **A205**, 541 (1951)
36. F.W. Bobrowicz and W.A. Goddard III (1977) The Self-Consistent Field Equations for Generalized Valence Bond and Open-Shell Hartree-Fock Wavefunctions. In: H.F. Schaefer III (ed) Methods of Electronic Structure Theory, vol 3. Plenum Press, New York, London
37. A.C. Wahl and G. Das (1977) The Multiconfiguration Self-Consistent-Field Method. In: H.F. Schaefer III (ed) Methods of Electronic Structure Theory, vol 3. Plenum Press, New York, London
38. J.A. Pople and R.K. Nesbet, J. Chem. Phys. **22**, 571 (1954)
39. C. Møller and M.S. Plesset, Phys. Rev. **46**, 618 (1934)
40. P. Hohenberg and W. Kohn, Phys. Rev. **136**, B864 (1964)
41. R.O. Jones (1987) Molecular Calculations with the Density Functional Formalism. In: K.P. Lawley (ed) ab initio Methods in Quantum Chemistry – Part I, Wiley, Chichester, New York
42. K. Balasubramanian and K.S. Pitzer (1987) Relativistic Quantum Chemistry. In: K.P. Lawley(ed) Ab initio Methods in Quantum Chemistry – Part I, Wiley, Chichester, New York
43. K. Fukui: Theory of orientation and stereoselection. Springer Verlag (1975), New York
44. R. Bonaccorsi, C. Petrongolo, E. Scrocco and J. Tomasi, Theor. Chim. Acta **20**, 331 (1971)
45. W.G. Richards: Quantum Pharmacology. Butterworths (1977), London-Boston
46. W.J. Hehre, L. Radom, P.v.R. Schleyer and J.A. Pople: *Ab initio* Molecular Orbital Theory. Wiley-Interscience (1986), New York
47. P.W. Atkins: QUANTA, A Handbook of Concepts. Oxford University Press (1991), Oxford
48. A. Hinchcliffe: Computational Quantum Chemistry. Wiley (1988), Chichester
49. A. Szabo and N.S. Ostlund: Modern Quantum Chemistry, Introduction to Advanced Electronic Structure Theory. McGraw-Hill (1989), New York
50. R. McWeeny: Methods of Molecular Quantum Mechanics. Academic Press (1989), London

P.E.M. SIEGBAHN AND MARGARETA R.A. BLOMBERG

OXIDATIVE ADDITION REACTIONS

1. Introduction

Two-electron oxidative addition reactions are defined by equation (1).

$$\text{(1)} \qquad M + \begin{matrix} A \\ | \\ B \end{matrix} \rightarrow M \begin{matrix} \diagup A \\ \diagdown B \end{matrix}$$

In this reaction where the A-B bond is activated, the metal center M is formally oxidized by two units as the two new M-A and M-B bonds are formed. The reverse reaction, the formation of the A-B bond, is termed reductive elimination. These two elementary reactions are common in organometallic reactions and they constitute significant steps in many catalytic cycles. An important example of reaction *1* is the oxidative addition of the H_2 molecule, which is an obligatory step in many catalytic cycles, such as hydrogenation of unsaturated hydrocarbons, hydroformylation and the reductive oligomerization of carbon monoxide [1]. The similarity between the H-H and the alkane C-H bonds, which are about equal in bond strengths and polarities, suggest that the oxidative addition might be a possible route for the activation of alkanes. Since alkanes are so abundant, they are very attractive as raw material for the synthesis of more useful organic molecules and therefore the search for effective and selective methods to activate this unreactive class of molecules have had high priority in chemical research for many years. A large number of theoretical studies have also been performed to elucidate the mechanisms of alkane activation. In this chapter a review is given of the most recent results obtained from studies of oxidative addition to transition metal centers. The review will focus on the activation of alkanes, but several departures will be made to gain a deeper insight into the mechanisms of the oxidative addition reaction. Subjects that will be covered here are, for example, the comparison between the activation of alkane C-H and C-C bonds with the activation of the H-H bond in the hydrogen molecule, elucidating the most basic reaction mechanism for the oxidative addition to a transition metal center. Furthermore, comparisons between alkane C-H activation and C-H activation in unsaturated hydrocarbons, together with comparisons between

P.W.N.M. van Leeuwen et al. (eds), Theoretical aspects of homogeneous catalysis, 15–63.

unstrained and strained C-C bond activation will give a more detailed understanding of the reaction mechanisms. A comparison of the oxidative addition of bonds with a varying degree of polarity, by the study of C-H, N-H and O-H activation will, among other things, show the effects of ligand lone pairs on the reaction mechanism.

Experimentally the first observation of the oxidative addition of an arene C-H bond to a metal complex was made by Chatt in 1965 [2]. Since C-H bonds in unsaturated hydrocarbons are known to be stronger than C-H bonds in saturated alkanes an intense research was started to find transition metal complexes which would activate the weaker alkane C-H bonds. However, even though many different metal complexes were found to activate arene C-H bonds, for years, the only type of alkyl C-H bonds that could be oxidatively added to metal complexes were intramolecular C-H bonds, with the alkyl chain connected to the metal by an intermediate ligating atom [3]. The first observations of alkane C-H insertion in solution, leading to moderately stable, isolable alkyl hydrides, were made in 1982 for iridium complexes, where the active intermediates were believed to be coordinatively unsaturated fragments of the general formula Cp^*IrL (L = CO,PR_3) [4, 5]. Shortly afterwards the analogous rhodium fragment (Cp^*RhL) was also found to be active [6] and later on also the $ClRhL_2$ (L = PPh_3) fragment [7]. Other metals proved to be effective in alkane oxidative addition by metal complexes are iron, rhenium and osmium [8]. Furthermore, the Cp^*ML (M = Ir,Rh) and CpML (M = Ir,Rh) fragments have been shown to insert into methane and other alkane C-H bonds in matrices at low temperature (10–20 K) [9–11]. These and other experimental results show an important trend, namely that oxidative addition reactions occur preferentially for metals to the right in the periodic table. In particular, rhodium and iridium seem to be the most effective metals for C-H activation by oxidative addition and overall very few metals have been observed to be active in this reaction. These observations imply that very special electronic structure requirements have to be fulfilled if the metal should activate alkanes. It is therefore of high interest to find out the main factors determining the differences between the metals in their abilities for oxidative addition reactions of different types.

Several explanations have been suggested for the fact that activation of the strong C-H bonds of arenes is much easier to observe than activation of the weaker C-H bonds of alkanes. One leading theory was that the π-coordinated precursor for the arene reactions was an important factor which should not be present in the alkane activation, and the former reactions should therefore have a kinetic advantage over the latter. In another theory the alkane oxidative addition reactions were assumed to be thermodynamically unfavourable and that all metal alkyl hydride complexes would be unstable to alkane reductive elimination [3, 12]. In fact, the experimental results for activation of different alkanes together with the arene results give rise to the surprising general trend that stronger C-H bonds are more easily activated than the weaker ones. Turning to the vinylic C-H bonds of alkenes, they are of intermediate strength to the arene and alkane C-H bonds. Furthermore, the alkenes, in contrast to the alkanes should have the same possibility as the arenes to form π-coordinated precursors. Still, C-H activation in alkenes was discovered later than both arene and alkane activation. It was not until 1985, that the first observation was made of the insertion of a mononuclear transition metal complex into the C-H bond of an unactivated alkene [13]. The reactive intermediate (η^5-C_5Me_5)(PMe_3)Ir was observed to form both a π-complex and a C-H oxidative addition product on reaction with ethylene [13, 14]. In the same study it was further shown that, for this particular system, the π-coordination of ethylene to the iridium complex is not a precursor to the C-H insertion product, but instead the two processes have to occur via independent transition states. This result is thus at variance with the arene activation result, where the π-coordination is a precursor for the C-H insertion step [3]. Later on a few more mononuclear transition metal complexes have been found to insert into unactivated alkene C-H bonds [15]. In all cases the olefin π-complex is believed to be the thermodynamically most stable product, while the vinylic C-H activation is the kinetically preferred reaction, due to the sterically demanding ligands in the complexes [14]. Recently, Jones and coworkers, in their studies of arene C-H activation by rhodium complexes, showed that the π-coordination of the arene in the precursor complex should not be too strong if the subsequent C-H activation reaction should occur [16]. Thus, for unsaturated hydrocarbons the formation of a π-complex with the metal can be an advantage due to kinetic factors or, due to thermodynamic factors, there can be a competition between π-coordination and the C-H insertion reaction. The questions of what role the π-coordinated complexes play for the oxidative addition reaction and why the stronger C-H bonds appear to be more easily activated than the weaker ones are thus also important questions to answer.

The direct intermolecular breaking of unstrained C-C bonds has still not been observed for any transition metal complex, even though this is a well known step in the breakdown of hydrocarbons on many transition metal surfaces. It is in this context interesting to note that both the H-H bond in the hydrogen molecule and C-H bonds are stronger by more than 10 kcal/mol than a C-C single bond. However, there are two classes of C-C activation reactions which have been observed in solution. First, intramolecular C-C activation reactions have been observed, in which a C-C bond in a coordinated ligand is cleaved. In these C-C activation reactions the product is stabilized by an increased unsaturation of the ligand. Secondly, the only type of intermolecular C-C activation by transition metals observed in solution involve C-C bonds with strain introduced by a ring structure of the carbon skeleton, such as in cyclopropanes or cyclobutanes. In particular the C-C bond in cyclopropanes is found to be activated by several transition metal complexes in solution. The first observation of this kind was made by Tipper in 1955 [17], who found that $PtCl_2$ reacted with cyclopropane to give a product with the formula $[PtCl_2 \times C_3H_6]_2$. The platinacyclobutane structure of the PtC_3H_6 unit was not elucidated until 1961, by Chatt and coworkers [18]. Later on many C-C activation reactions of strained alkanes by transition metal complexes have been observed, involving most commonly rhodium and palladium complexes, see for example the review article by Crabtree [19] and references therein. It should also be mentioned that nickel atoms have been shown to spontaneously insert into the C-C bond of cyclopropane in matrix isolation studies [20]. The fact that no unstrained C-C bonds have been observed to insert intermolecularly to transition metal complexes together with the above mentioned results for C-H and H-H activation raises the important question of why H-H activation is easier than C-H activation, which in turn is easier than C-C activation. Furthermore, the fact that cyclopropanes are more easily activated than cyclobutanes, in spite of the similarity in C-C bond strengths, leads to the question of precisely how the strain in the C-C bonds affects the reaction.

The experimental results discussed above are mainly concerned with the reactivity of metal complexes in solution. Also the gas phase reactivity of naked metal atoms has been studied experimentally. Recently measurements of reactivities of neutral metal atoms have been performed [21–24], whereas earlier studies were mainly concerned with the reactivities of metal cations. The reactivity of transition metal cations with hydrocarbons has been studied intensively during the last decade [24–27]. One obvious difference between the neutrals and the ions is that the metal ion has a long-range interaction with the polarizable hydrocarbon through the charge-induced dipole interaction. This attraction has been thought to be strong enough to overcome the barrier for C-H bond activation that hinders the reactions of the neutral systems [25–27]. Thus, the oxidative addition of C-H bonds has been expected to be more facile for cations than for neutral metal atoms. However, recent experiments for neutral atoms [24] indicate that this difference might not be as large as thought only a few years ago. In this review the differences in the mechanisms for the activation of alkanes between cations and neutral systems will also be discussed.

If there is any hope for major contributions to chemistry from theory, the conclusions drawn must have some generality. The approach which appears best chosen to observe general effects is the study of trends. As will be seen in the method section, this is also the approach which tends to reduce the importance of having to do very accurate calculations on each individual system. The relative accuracy between different systems is much higher than the absolute accuracy obtainable with the present type of methods. The choice of systems to study is then reduced to finding suitable general problems where the answers to the questions are of high interest and yet the problems should not be so trivial that the answers to the questions are obvious without any calculations. In the presentation of the experimental background above, we have tried to identify a few such general questions. The approach taken by our research group is thus to study trends of different kinds [28–42] and this type of studies are the ones that will be mainly discussed in this chapter. To study trends means to study many systems and for this to be possible the systems must not be too large, i.e. simple model systems are preferably used rather than large, realistic metal complexes. Another obvious reason for this choice is that the simpler the model system is the easier it is to identify the properties of main importance for the particular problem of interest. The simplest model of this kind for a transition metal complex is the naked metal atom, and all our studies therefore take the starting point in the reactions of naked metal atoms. Ten years ago, many chemists would argue that no results of general chemical interest can emerge from a study of a model of transition metal complexes without ligands. However, one of the goals of the present review is to show that much more interesting conclusions can be drawn from these simple model systems than was originally believed a decade ago. It should furthermore be made clear that, the opposite result where the transition metal atoms were found to behave completely differently from normal transition metal complexes, would also have been a very useful result. In fact as will be shown in Section 8 below, there is no problem in the present approach to then proceed from these atomic models and add ligands in a systematic way to approach more realistic models.

One simple way to obtain a trend of results is to study an entire row of transition metals for a given problem. Most of the problems discussed in this review are of such general character that any of the three transition metal rows would suffice for this purpose. As will be shown in the method section below, the second row is easiest to treat and this row was therefore chosen for most of the studies discussed in this review. The main advantage with the selection of an entire row for each problem is that a representative group of different systems are compared. It has been demonstrated in previous studies by Bauschlicher, Langhoff and coworkers [43] that this is a very useful approach to obtain a quantitative analysis of the energetics in reactions involving transition metals. For example, this approach allows for a systematic evaluation of the importance of the positions of the various atomic states on the metal atom for the bonding. Also, the effects of a continuous increase of the ionization potential and an increase of the number of d-electrons of the metal can be investigated by going from left to right across the row in the periodic table. A similar approach to study trends has also been taken by Balasubramanian and coworkers, who have extensively investigated the oxidative addition of all second row transition metal atoms to H_2 [44].

One of the main conclusions that has been possible to draw on the basis of the results obtained from studies of a whole row of transition metal atoms, concerns the loss of exchange energy [26, 28, 43, 45] in the reaction. This energy loss is particularly large for the atoms in the middle of the row since there is a large number of unpaired 4*d*-electrons for these atoms. Therefore, the binding energies between naked metal atoms and practically any ligand will display a marked minimum in the middle of the row.

The question concerning differences in the carbon-hydrogen and the carbon-carbon activation mechanisms has been addressed previously in model calculations. In a study comparing alkane activation to the activation of molecular hydrogen [46, 47] it was found that the alkane activation reactions have higher barriers than the hydrogen activation reaction. This difference was suggested to be due to the directed nature of the methyl group compared to the spherical nature of the hydrogen atom. This directionality makes it energetically more unfavourable for methyl groups than for hydrogen atoms to bind both to the metal and the other R group in the transition state structure. Thus, a C-C activation barrier of 42 kcal/mol was obtained in that study [47], compared to 3 kcal/mol for the H-H activation barrier. Later on also the correct order between the C-H and C-C activation barriers was obtained by Low and Goddard [48], in a study on the palladium and platinum reactions with ethane and methane, where, for example, the barrier height for C-H activation by palladium was calculated to be 30 kcal/mol, compared to 39 kcal/mol for the C-C activation.

Two recent reviews of theoretical studies of the oxidative addition reaction to metal centers have appeared, one is written by Koga and Morokuma [49] and the other is written by Hay [50]. These reviews describe results of the more traditional approach where calculations are performed on a few realistic systems, in order to make direct comparisons to experimental findings. Kitaura, Obara and Morokuma have studied the oxidative addition of H_2 to $Pt(PH_3)_2$ and determined the three-centered transition state using energy gradient methods [51]. This was the first theoretical determination of the transition state structure for an organotransition-metal reaction [49]. The same reaction was studied by Noell and Hay [52]. Koga and Morokuma [53] have investigated the details of the C-H activation in methane using the model complex $ClRh(PH_3)_2$. They showed that this complex dissociates the C-H bond of methane without any barrier. Methane activation by the oxidative addition mechanism has also been studied by Ziegler and coworkers, who investigated the methane activation by the CpML complex, with M = Ir,Rh and L = CO,PH_3 and also by $M(CO)_4$ complexes for M = Ru,Os [54]. One of the results in that study is that the complex RhCp(CO) has only a very small barrier for the methane activation. Both these studies also found important precursors with molecularly bound methane.

2. Computational methods

The decision on computational strategy for a given problem is often both the most difficult and the most important part of a quantum chemical investigation. If the problem is well chosen from a chemical viewpoint, it is normally a risk to choose a strategy which is either too accurate or one which is not accurate enough. It is clear that if the computational strategy chosen does not turn out to be accurate enough for the problem, the outcome of the calculations will tend to be rather uninteresting. On the other hand, if the chosen method is too accurate it is in general also costly, which means that it can put a restriction on either the size of system or the number of systems to be studied, which can severely restrict the interest of the calculations performed.

For the present type of studies of transition metal complexes the problem of choosing a computational strategy can be divided into two parts. First, a level of computation has to be selected for the geometry optimization. Secondly, methods and level of accuracy have to be selected for the calculation of the energies at the optimized geometries. These two choices should be made in a balanced way so that not unnecessary time is spent in a section which is not decisive for the final overall accuracy. Precisely how the optimal strategy should be chosen depends on the systems of interest. For example, the optimal strategy to treat transition metal complexes is different for each of the three rows in the periodic table. For complexes of the first row of the transition metals, the geometry optimization is normally much more difficult and more sensitive to the level of treatment than it is for the other two rows. On the other hand, the treatment of relativistic effects will be more difficult the heavier the elements are, and for the third transition row a treatment without inclusion of spin-orbit effects is questionable. Overall, the presently available quantum chemical *ab initio* methods are best suited to treat the second row of the transition metals. In the present review we will therefore focus on the treatment of second row complexes but a few examples will also be given from studies of first row complexes.

2.1. GEOMETRY OPTIMIZATION

The geometry optimization step is an essential part of every modern study of transition metal complexes. Efficient and general program systems are available and the problem is just to choose the proper level of optimization. At least four different main types of treatment are of interest in this context. First, at the lowest level, there is the standard Self-Consistent Field (SCF) method. At the second level some degree of electron correlation is introduced using a Multiconfigurational SCF (MCSCF) scheme. The most common variant of this method is the Complete Active Space SCF (CASSCF) method. This treatment mainly takes care of near-degeneracy effects. The simplest treatment of dynamical correlation effects is the Møller-Plesset second order perturbation theory (MP2). Finally, at the highest level of treatment there are several choices but we will here just mention two of these methods. The first of these methods are the Coupled Pair Functional (CPF) [55] or Modified CPF (MCPF) [56] methods. The second of these methods mentioned here belongs to the Coupled Cluster Methods and is called the Quadratic Configuration Interaction Singles and Doubles (QCISD) method [57]. The choice of level for the geometry optimization is further complicated by the choice of basis sets. Two different choices of basis sets will be discussed here. First there are the standard double zeta basis sets and secondly there are basis sets including polarization functions. For the cores of the transition metals, standard relativistic Effective Core Potentials (ECP's) [58] can be used without loss of accuracy.

So far, the choice of geometry optimization has been based on tests on selected individual systems of interest and some of these tests will be mentioned below. However, a more systematic investigation of the accuracy of

TABLE I

Calculated MCPF energy differences (kcal/mol), relative to the energy obtained for the SCF structure, for selected second row transition metal complexes. In the different columns the results using different levels of geometry optimization are given. A positive energy difference means that the lowest energy is obtained for the SCF structure

System	SCF	MP2	QCISD
YH_2CH_3	0.0	+0.1	+0.4
ZrH_3CH_3	0.0	+0.4	+0.8
NbH_4CH_3	0.0	+0.3	+0.8
TcH_4CH_3	0.0	+1.2	+1.4
RuH_3CH_3	0.0	+0.9	+1.4
RhH_2CH_3	0.0	+1.8	+1.4
$PdHCH_3$	0.0	+1.1	+0.8

the above mentioned geometry optimization schemes is under way [59] and the preliminary results of that study will first be given here. The test systems for that study were taken from a previous study of the oxidative addition of methane to second row transition metal hydrides [32]. Going from left to right in the periodic table, the following closed shell equilibrium structures were chosen, YH_2CH_3, ZrH_3CH_3, NbH_4CH_3, TcH_4CH_3, RuH_3CH_3, RhH_2CH_3 and $PdHCH_3$. The corresponding molybdenum complex MoH_5CH_3 was left out since the presence of multiple minima complicated the analysis. To compare the different geometry optimization methods, the total energy was calculated for each of the optimized structures using the same method. This is the MCPF method including relativistic effects using basis sets with polarization functions. The preliminary results of this comparative study are shown in Table *I*. The energies in the table are given with the results for the SCF geometries as reference energies. Standard double zeta basis sets were used in the geometry optimization in all cases. As seen from the results in the table, the total energies calculated using the different geometries are very similar. In fact, the best (the lowest) energies are in all cases obtained for the SCF geometries. This is of course to some extent due to the choice of basis sets. Better basis sets including polarization functions in the geometry optimization, were tried for the yttrium and the palladium complexes and with these basis sets the energies at the QCISD geometries are slightly lower (by 0.1–0.3 kcal/mol) than the energies at the SCF geometries, as they should be. It should be noted that the entries in the table relate to total energies. For relative energies, such as binding energies, the results obtained using the different methods are even more similar. Considering that the geometry optimization is by far fastest in the SCF case, the choice of method for the geometry optimization of second row transition metal complexes is easy. The SCF level is optimal.

There are several other investigations of geometry optimization methods for second row transition metal complexes, which point in the same direction. Most noteworthy of these are the ones by Bauschlicher and coworkers, who generally conclude that if a consistent set of ligand and metal-ligand geometries is used, the binding energies calculated at the MCPF level agree to better than 1 kcal/mol, regardless of whether the equilibrium structures are optimized at the SCF or MCPF level of theory [43]. The situation for transition state structures is probably similar but has been less tested. However, a few tests are available. For example, it was shown in Ref. 28 for the methane activation reaction that the barrier height for rhodium, the metal with the largest correlation effects in the oxidative addition reaction, changed by less than 1 kcal/mol on going from an SCF- to an MCPF-optimized geometry. Also, a preliminary test of the barrier height for the oxidative addition of water to the palladium atom, comparing the three methods in Table *I*, gives the same result.

The origin of the rather surprising result of the high accuracy for the SCF optimized results, is that in the most interesting region of the potential energy surfaces (including both the transition state and the insertion products) the SCF and highly correlated surfaces are quite parallel. This is seen on the rather small correlation effects on the elimination barriers. For example, for the barrier of ethylene elimination from palladium-vinyl-hydride [31] the SCF and the MCPF values are identical, and for the corresponding rhodium reaction the correlation effects lower the elimination barrier by only 4 kcal/mol, compared to 56 kcal/mol for the activation barrier. Another reason SCF geometries can be used is that the potential energy surfaces are often rather flat in both the transition state region and the insertion product region, so that discrepancies in SCF- and correlation-optimized structures have very small effects on the relative energies. The conclusion is that the use of SCF-optimized structures normally give reliable results for the trends in activation energies and binding energies for second row transition metal complexes.

There are a few exceptions to the general rule given above and these cases are easy to identify. These are the cases where the complex can dissociate properly at the SCF level. For example, the metal-ethylene distance in $Pd(C_2H_4)$ becomes much too long at the SCF level [29]. For some of these systems the MP2 method may work better but this is not always the case

TABLE II
Calculated MCPF energies, relative to the triplet ground state asymptote, for the NiH(OH) $^1A'$ insertion product and transition state. The optimization method concerns the level of geometry optimization

Optimization method	DE_{IP} [kcal/mol]	DE_{TS} [kcal/mol]
SCF	+ 1.3	+ 9.0
MP2	−12.6	+11.4
CASSCF	−11.0	+ 9.8
QCISD	–	+ 2.9
MCPF	−13.2	+ 3.1

[57], so that in these situations a choice of geometry based on experimental results or a high level correlation treatment is recommended.

All the above mentioned examples were taken from the second transition metal row. The situation for the other two rows is less well known. However, a few tests have been made. The results in Table *II* are taken from a study of the reaction between water and the nickel atom [60]. The results in this table indicate that the final energies are much more sensitive to the geometry optimization in this case than for the above mentioned second row systems. This is true, in particular, for the transition state where only the highest level treatments are sufficient. It should be added that the highest level results are in very good agreement with experiments. The conclusion that the geometry optimization is more difficult for first row than for second row systems is in line with other experience for similar systems. The origin of this difference between the rows is that the sizes of the valence d- and s-orbitals are much more different for the first row than they are for the second row. This leads to much weaker d-bonds for the first row with less optimal overlap, which in turn leads to the normal near-degeneracy problems that require a multiconfigurational treatment. This type of problem should not appear for the third row elements since the sizes of the d-and s-orbitals are similar, and for complexes of this row an SCF geometry optimization might therefore be sufficient.

2.2. ENERGY CALCULATION

For the present type of transition metal systems, the limitation of the accuracy will always be in the final energy evaluation at the optimized geometries. It is therefore of large importance to be as accurate as possible in this step of the calculation. The dynamical correlation energy of all the valence electrons of the system has to be obtained using relatively large basis sets. To be able to obtain a sufficiently large fraction of the valence correlation energy, basis sets including polarization functions are needed, possibly with the exception for some ligands far away from the reaction center. The inclusion of at least one set of f-functions on the metal is also to be recommended. The question whether it is possible to obtain useful results if only a limited number of the valence electrons is correlated was approached for the $Ni(CO)_x$ systems a few years ago with negative results [61]. For example, if only the metal $3d$,$4s$ valence electrons and the two carbon lone-pair electrons are correlated, only about 70% of the binding energy of NiCO was obtained. If some of the valence electrons should be left out it is at least clear that a rather sophisticated localization scheme is required.

The choice of method for the correlation treatment is of large importance for the final accuracy. Since this is the accuracy limiting step it is to be recommended that the most accurate methods available are used. At present this means that some type of coupled cluster method should be used. Multireference CI methods are obviously also adequate but this choice will normally severely limit the size and the number of systems to be studied. Of the coupled cluster type methods it is probably not very important which choice is made. The CCSD (Coupled Cluster Singles and Doubles) method or the above mentioned QCISD methods are of about equal accuracy. The CPF or MCPF methods are somewhat simpler variants of the coupled cluster methods but experience has shown that the results of these methods are in general at least as high as for the more rigorous methods. The MCPF method has therefore been used in most of the applications discussed in this paper. The main limitation of these coupled cluster methods is that the effect of triples is not included. General methods of high efficiency where the triples are included will soon be available [62] and these methods will then be of high interest to use on transition metal complexes. Of the somewhat simpler correlation methods the MP2 method has been used frequently on transition metal complexes [49]. Even though this method can generate useful results, this method has to be used with some care for transition metals. It has, for example, been shown that the MP2 method yields a splitting between the lowest states of the nickel atom that is in worse agreement with experiments than the results obtained at the SCF level [63]. Since these splittings often are of large importance for the energetics, the use of the MP2 method for complexes of the nickel atom and other related atoms, can not be recommended in general.

It was mentioned above that near-degeneracy effects can be significant for transition metal complexes, in particular, for first row transition metals. Ten years ago, the presence of coefficients in the MCSCF expansion larger than 0.10 would have lead to the conclusion that an MCSCF treatment followed by a multireference CI would be absolutely necessary. However, one of the most surprising findings in the computations done during the last decade is that single reference state methods of the coupled cluster (or MCPF) type, are able to handle even rather severe near-degeneracies quite accurately. This is a major computational advantage since the single reference state methods are much faster and also much easier to use routinely. When it will be possible to include also the triples in the coupled cluster methods, it will probably even be difficult to find transition metal complexes for which the presence of near degeneracy effects is a dominating problem, at least in comparison to the basis set problem. It is not unlikely that the well-known chromium dimer will be an almost unique such case.

To the valence electrons of a second row transition metal are normally counted only the *4d*- and the *5s*,*5p*-electrons, and correspondingly for first and third row transition metals. The problem of leaving out the metal *4s*,*4p*-electrons from the correlation treatment is not always trivial and is in our opinion a severely underestimated problem. For example, with the large correlation energy of the *4d*-shell for the atoms to the right, even a slight mixing between the *4p* and *4d* orbitals can cause large changes in the computed valence correlation energy. Also, with halogens or oxygens present among the ligands there can be large rotations between the metal *4s*,*4p*-orbitals and the inner valence orbitals of the ligands. This problem could in principle be solved by also correlating the *4s*,*4p* electrons of the metal atoms but this would make the calculations much more expensive, partly because this would require a significant increase of the basis sets in the *4s*,*4p* region. Furthermore, based on our experience, this type of rotation does not indicate that it should be important to correlate the metal *4s*,*4p*-electrons. However, if the *4s*,*4p* electrons are left out of the correlation treatment, a localization of the metal *4s*,*4p*-orbitals is in general required. In this context we have found a method based on a minimization of $< r^2 >$ of the core orbitals useful [64], but simpler methods based on maximization of certain coefficients in the core orbitals have also been used with satisfactory results.

When calculations on transition metal complexes are made, relativistic effects have to be included. The mass-velocity and Darwin terms can be trivially included in the calculations since these terms are one-electron properties [65]. For first and second row transition metal complexes this turns out to be a sufficient treatment of relativistic effects in most cases. For third row transition metal complexes the situation is less clear. Even though there are many cases where spin-orbit effects are less important there are a large number of situations where these effects can be quite significant. Some estimate of spin-orbit effects should therefore ideally be made when third row complexes are treated. This can be done using effective one-electron operators or more sophisticated methods [44].

2.3. ACCURACY OF THE RESULTS

When the above mentioned quantum chemical methods are used on transition metal complexes, a very high absolute accuracy can not be expected in general. This is clear already from an inspection of the results obtained with these methods for simple molecules like the hydrogen molecule and methane. For the binding energy of H_2 a value of 105.6 kcal/mol is obtained using the MCPF method with a basis set including only one p-function on hydrogen. This result is too low by 3 kcal/mol compared to the exact result. Similarly, for methane the error in the C-H bond energy using a double zeta basis with polarization functions is 5 kcal/mol. Basis sets much larger than these on hydrogen and carbon are not realistic to use on the transition metal complexes discussed in this paper. Based on these simple comparisons, it is clear that absolute errors of 7–8 kcal/mol must be expected for the bond strengths of transition metal-hydrogen and transition metal-carbon single bonds, if a basis set including only one f-set is chosen on the metal and only one polarization function is used on carbon and hydrogen.

In connection with estimating the absolute accuracy of the calculations on transition metal complexes, available experimental results mainly for cationic systems are very useful [24–27]. Several comparative studies for cationic systems have therefore been made [43, 66]. These comparisons have essentially confirmed the above general expectations. If double zeta plus polarization basis sets are used for first row atoms and a basis set with a triple zeta description of the valence *d*-orbital and one f-set on the metal is used, an absolute error of 7–8 kcal/mol for single bonds to the metal is found at the MCPF level. Bauschlicher *et al.* [67] have shown that for smaller systems it is possible to obtain significantly higher absolute accuracy already today. However, results at different levels of treatment for methane indicate that substantial increases of the basis sets are needed to significantly increase the absolute accuracy. Using very large s,p-basis sets including d-functions on hydrogen and f-functions on carbon still give errors on the C-H bond strength in methane of 2 kcal/mol at the MCPF level [67]. Therefore, for the near future, the most significant increase of the absolute accuracy that can be expected will rather come from the use of methods including triples than from the use of larger basis sets.

It is important to note that the above discussion concerned the *absolute* accuracy of the calculations. One of the most important properties of *ab initio* methods is that the errors are always systematic. For example, bond strengths are always underestimated. In fact, if the errors would have been random very few results of chemical use would have been obtained with these methods. Therefore, much more useful information is obtained if trends of results are studied. Fortunately, trends of results are also in general of much larger chemical interest than results for separate individual systems, even if the accuracy would be higher for these latter systems. In the rest of this paper the discussions will therefore be focussed on trends. If absolute energies are of interest, even the above rather crude general estimates of the bond strength errors are quite useful. Applying corrections of +3 kcal/mol for the H-H bond, +5 kcal/mol for a C-H bond and +8 kcal/mol for a single M-C or M-H bond, gives very good quantitative agreement with available experimental information for cationic systems. At the same time these corrections leave the trends between the different metals entirely unchanged.

2.4. COMMENTS ON SPIN-STATES

Finally, a general and rather complex question will be addressed. For most of the reactions discussed in this review, the ground state of the reactants normally has a different total spin than the ground state of the products. Two comments can be made in this context. First, the question whether the binding energies should be given relative to reactants with the same spin as the products or relative to the spin of the ground state reactants is mainly a pedagogical problem. One set of energies can be easily transferred to the other set using available excitation energies. The common praxis has been to relate to the energies for the ground spin states of the reactants, and this praxis will be followed here. The main advantage with this praxis is that the procedure is well defined. A more serious question concerning the spin states is what actually happens dynamically during the reaction. If the reaction starts with ground state reactants and ends up with ground state products with a different spin, the spin has to change through spin-orbit effects. These effects are known to be strong for transition metals so this surface-hopping is intuitively expected to occur with a high probability. This problem has been studied in detail by Mitchell [68], who showed that in the case of the association reaction between the nickel atom and carbon monoxide, the crossing probability is near unity. Also, in order to rationalize the experimental results for the oxidative addition reaction between the nickel atom and water, a high crossing probability has to be assumed [60]. Since the potential surface for the high spin reactants is normally strongly repulsive, the crossing between the two spin surfaces will in most cases occur far out in the reactant channel, long before the saddle point of the reaction is reached. This is at least true in the most interesting cases where the low spin surface of the reactants is not too highly excited. For the oxidative addition reaction this assumption is, for example, confirmed by calculations by Balasubramanian [44]. This means that the probability for surface-hopping through spin-orbit coupling will affect the pre-exponential factor of the rate constant, but not the size of the barrier. The computed barrier heights discussed here should therefore in most cases be directly comparable to experimental measurements of activation energies.

3. Activation of H-H, C-H and C-C bonds

In this and in the following sections, general questions related to the oxidative addition reaction will be discussed. As mentioned in the introduction, the approach which appears best chosen to observe general effects is the study of trends. One general problem well suited for this type of approach is the question why the ease of activation decreases in the sequence of H-H, C-H and C-C bonds. In order to identify the metal properties of main importance a suitable set of systems should be selected. For this study and most of the other studies discussed in this review, the entire second row of the transition metals was chosen to be this set of systems. For such a general problem as the activation of H-H, C-H and C-C bonds, any row of the transition metals could have been chosen for the study, but from the conclusions in the preceding section it is clear that the second row is easiest to treat. The analysis of the effects will start with the results obtained for the reactivity of the naked metal atoms. The effects of ligands will be discussed in a section below. A general question based on experimental results, which will also be addressed in the discussion below, is why the activation of C-H and C-C bonds appears to be easier for the transition metals to the right. For the second row, this seems to be the case, in particular, for rhodium complexes.

The results for the reaction energies and barrier heights for the reactions between second row transition metal atoms and the hydrogen molecule, methane and ethane are given in Figures *1* and *2* [28, 30–32]. Since most of the

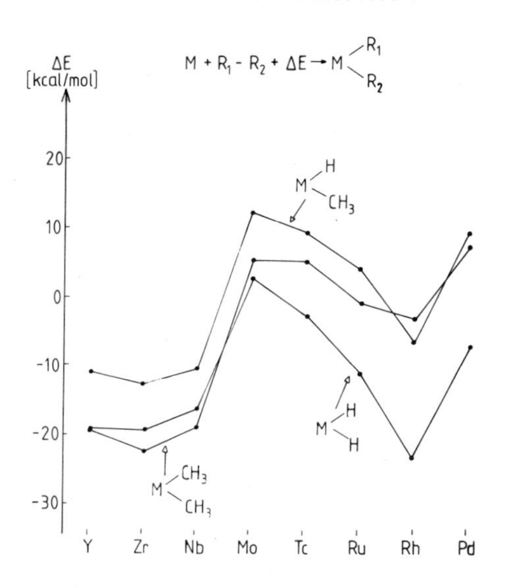

Fig. 1. Insertion product energies for the H-H, C-H and C-C activation of the hydrogen molecule, methane and ethane, respectively. The energies are calculated relative to ground state reactants.

reactions with the hydrogen molecule did not give any barriers on the low-spin surface, the H_2 results were left out in that figure. From the results in Figure *2*, it can be immediately seen that the results, even for these ligand free complexes, are consistent with general experimental information for ligated complexes. The barriers are lowest for the hydrogen molecule, followed by the C-H activation of methane and with much higher barriers for the C-C activation of ethane. Also, and perhaps more surprisingly, the C-H activation barrier is actually lowest for rhodium in agreement with the fact that only rhodium complexes in the second row have been found to activate C-H bonds of saturated hydrocarbons [8]. When the situation is as fortunate as it is here, it is a major advantage from the viewpoint of understanding the origin of the results, that the ligands have already been removed from these systems.

With these model results available, the main origin of the difference in the activation of H-H, C-H and C-C bonds is easy to understand. When the hydrogen molecule approaches a transition metal the bonds to the metal can start to form gradually as the H-H bond is weakened since the spherical hydrogen atom can bind in different directions. This usually leads to an oxidative

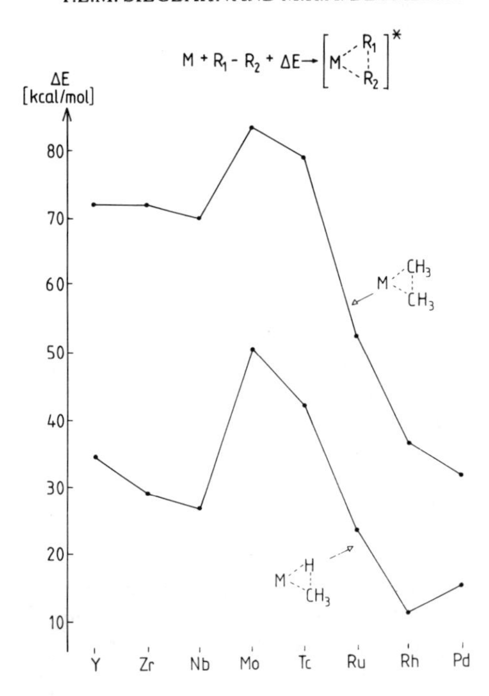

Fig. 2. Transition state energies for the C-H and C-C activation of methane and ethane, respectively. The energies are calculated relative to ground state reactants.

addition reaction with no or only a small barrier. The methyl group is different from hydrogen and forms more or less strongly directional bonds. Therefore, when a C-H bond or a C-C bond approaches a metal the methyl carbon can only start to efficiently bind towards the metal when the methyl group is tilted towards the metal. In this process the C-H or the C-C bond first has to partly break, which costs energy and leads to larger barriers than for the activation of H-H bonds. When two methyl groups have to tilt as in the activation of C-C bonds, a larger barrier is obtained than when only one methyl group is involved as in the breaking of C-H bonds. These trends have been clearly demonstrated by several quantum chemical calculations [48, 69].

An interesting observation concerning the energetics in the three different types of reactions can be made from the results in Figures *1* and *2*, which supports the above reaction mechanism. For most metals the barrier for alkane elimination, which is the reverse of the oxidative addition reaction, is about

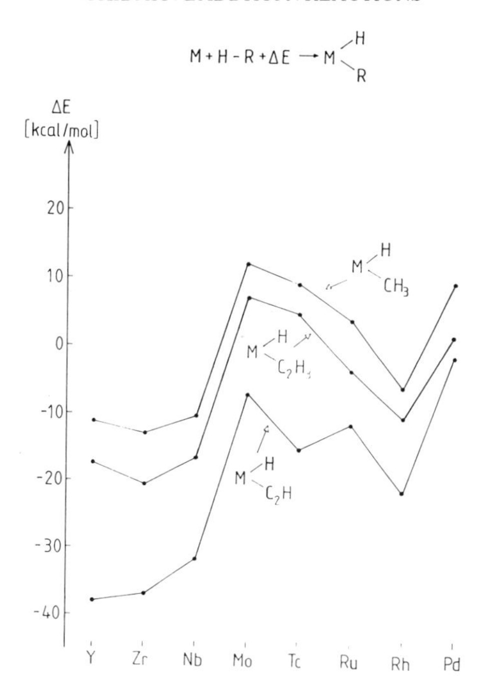

Fig. 3. Insertion product energies for the C-H activation of methane, ethylene and acetylene, respectively. The energies are calculated relative to ground state reactants.

twice as high for ethane as it is for methane. With the tilting of the methyl group as the barrier determining factor, this result is logical since two methyl groups are tilted for ethane and only one for methane. A similar simple relationship does not exist for the addition barriers, since other effects such as promotion and exchange loss effects also enter the size of these barriers. The relationship between the barrier heights for the C-H and C-C elimination reactions was first pointed out by Low and Goddard [48] for palladium. As a curiosity it can be noted that for this particular system, later more accurate calculations [30, 69] gave a much larger ratio of almost four between these elimination barriers. However, a ratio of two was obtained for most of the other metals.

Three main conclusions concerning the electronic structure aspects emerge from the results of these studies of the reaction between the naked transition metal atoms and hydrogen, methane and ethane. First, the main state involved

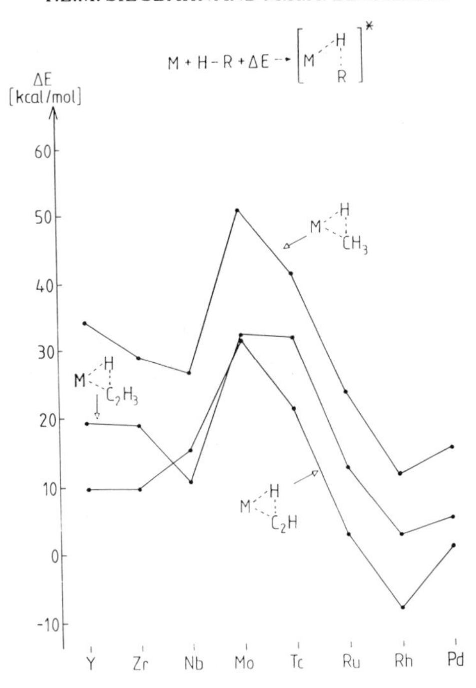

Fig. 4. Transition state energies for the C-H activation of methane, ethylene and acetylene, respectively. The energies are calculated relative to ground state reactants.

in the binding in the insertion products is the s^1-state (or longer, the $d^{n+1}s^1$-state). For the second row atoms to the left, there are also important contributions from s^1p^1-states (or longer, $d^ns^1p^1$-states). The second main conclusion is that at the transition state the s^0-state (or longer, the d^{n+2}-state) plays a key role. It is the presence of this low-lying state that leads to the lowest barriers for the atoms to the right, ruthenium, rhodium and palladium. In particular, the lowest barrier of the second row atoms is found for rhodium since this atom has both low-lying s^0- and s^1-states. The origin of the importance of the s^0-state in the transition state region is that this is the state with least repulsion towards ligands and therefore allows the metal to approach as close as possible for a more favourable interaction. The third main conclusion concerns the loss of exchange energy in the reaction. This energy loss is particularly large for the atoms in the middle of the row since there is a large number of unpaired 4*d*-electrons for these atoms. Therefore, the binding energies between naked metal atoms and practically any ligand will display a marked minimum in the middle of the row.

Another general finding in the above studies for the oxidative addition with methane, is that the metals to the right form relatively strongly bound molecular adducts with methane on the low-spin surface. However, since most of the atoms have a high-spin ground state these molecular complexes are not bound compared to the ground state atoms and methane. The only complex which is bound with respect to the ground state asymptote is palladium. The importance of the s^0-state for the binding in the molecular complex on the low-spin surface is even more pronounced than it is at the transition state region. The close correspondence between strongly bound molecular complexes and low barriers is also in line with general experimental findings.

4. C-H activation in unsaturated hydrocarbons

One mystifying and surprising result that has emerged from the active research on the C-H oxidative addition reaction is that there is a rather general anti-correlation between the initial C-H bond strength and the difficulty to activate this bond by transition metal complexes [3, 12], which is exactly opposite to the simplest expectations. C-H bonds considered in this experimentally found anti-correlation are the ones in methane, in ethylene, in benzene, in progressively larger alkanes and in substituted alkanes. In order to study this general anti-correlation, the activation of three different types of C-H bonds were investigated. The first of these is methane [28, 31], which was discussed already in the preceding section. The C-H bond strength in methane is 112 kcal/mol. The second C-H bond is the one in ethylene [31] with a bond strength of 118 kcal/mol and finally one of the strongest known C-H bonds, the C-H bond in acetylene [39], with a bond-energy of 140 kcal/mol was studied. The entire second row of naked transition metal atoms were studied. The results for the reaction energies are shown in Figure *3* and for the barrier heights in Figure *4*.

Of the three hydrocarbons acetylene, ethylene and methane, acetylene has the lowest barrier for C-H activation followed by ethylene and methane for almost all second row transition metal atoms. This trend is thus in line with the general experimental evidence of an anti-correlation between the initial C-H bond strength and the difficulty to activate this bond by transition metal complexes. The simplest explanation of the origin of the differences in barrier heights is that it is a steric effect. Since C-H activation requires that the metal efficiently interacts in a sideways orientation with the C-H bond, it is

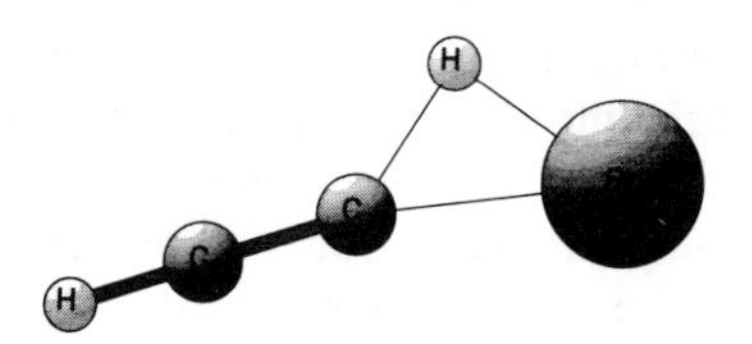

Fig. 5. Transition state structure for the C-H activation of acetylene by rhodium.

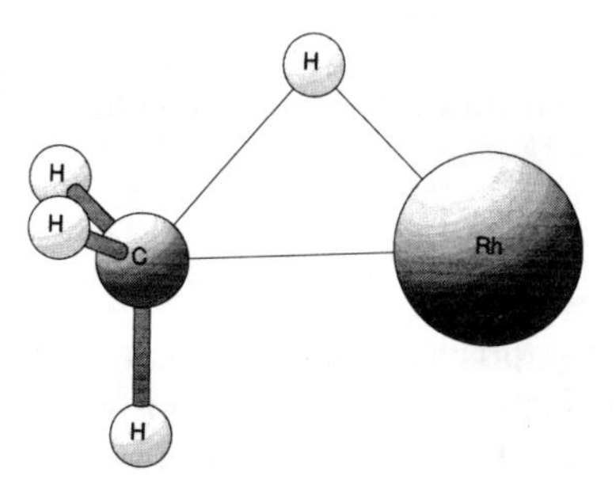

Fig. 6. Transition state structure for the C-H activation of methane by rhodium.

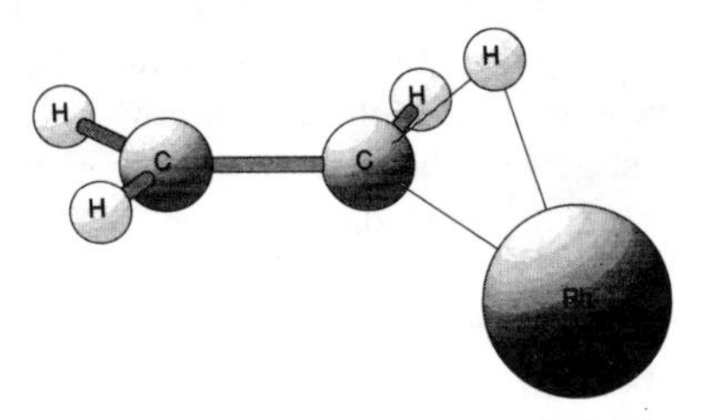

Fig. 7. Transition state structure for the C-H activation of ethylene by rhodium.

clear that this position is most easily reached for acetylene, see Figure *5*. For methane, on the other hand, a substantial initial distortion of the molecule is needed to reach a proper interaction, see Figure *6*. For ethylene the situation is somewhere in between that of acetylene and methane, see Figure *7*.

Another contributing factor of major importance for the order of the barriers is that the exothermicities of the reactions also follow the same order. The major origin of the differences in exothermicities is the interaction between the metal and the π- and π^*-orbitals of ethylene and acetylene. In particular, the donation from the π-orbitals to the metal is important and leads to larger differences in the exothermicities for the atoms to the left where there are empty 4*d*-orbitals. This effect is strongest for acetylene since there are two π-orbitals and for yttrium and zirconium which have two empty 4*d*-orbitals. The interaction with the π- and π^*-orbitals are special for ethylene and acetylene and does therefore not explain why the strongest C-H bonds are easiest to activate in general. This trend can instead be understood from the fact that the same factors that affect the C-H bond strength of the hydrocarbons are to some extent active also for the strength of the M-C bonds. These factors are both hybridization effects and steric repulsion effects. The C-H bond is strongest where the carbon atom has the least number of additional ligands, that is for acetylene. The same argument holds for the M-C bond strength but to an even larger extent due to the larger size of the metal atom, and this will thus lead to larger exothermicities for the insertion into the strongest C-H bonds.

A notable result in Figures *3* and *4* is that the behaviour of the reaction energies is quite similar for the three reactions. For example, the lowest barrier is in all cases found for rhodium. It is thus clear that the same electronic mechanisms, described in the previous section, are involved in the three cases. This means that it is of major importance for a low barrier to have both a low-lying s^0- and s^1-state for all three reactions.

A complicating factor in the activation of the C-H bonds of ethylene and acetylene is that these systems form strongly bound π-complexes. This means that even though the barriers for C-H insertion of acetylene are small for many metal atoms, actually absent for Ru, Rh and Pd, this reaction is not thermodynamically favoured in the gas phase without additional ligands for most metals. The only exception to these results for acetylene is technetium which actually has a more stable insertion product. However, at the same time technetium has one of the highest barriers for C-H insertion of about 20 kcal/mol. The possibility to observe the C-H activation for technetium is therefore not very high either. Also the alkene π-complexes are so strongly bound that they tend to be clearly thermodynamically favoured over the metal inserted vinyl-hydride complexes. This is actually true even for the complexes for which C-H activation of alkenes were finally observed experimentally. The reason C-H insertion could still be observed for these complexes is that this process is kinetically favoured. This was accomplished by blocking π-complex formation through the use of bulky, sterically demanding ligands on the metal. The present results for the second row transition metal atoms indicate that it might not be impossible to find complexes where the vinyl-

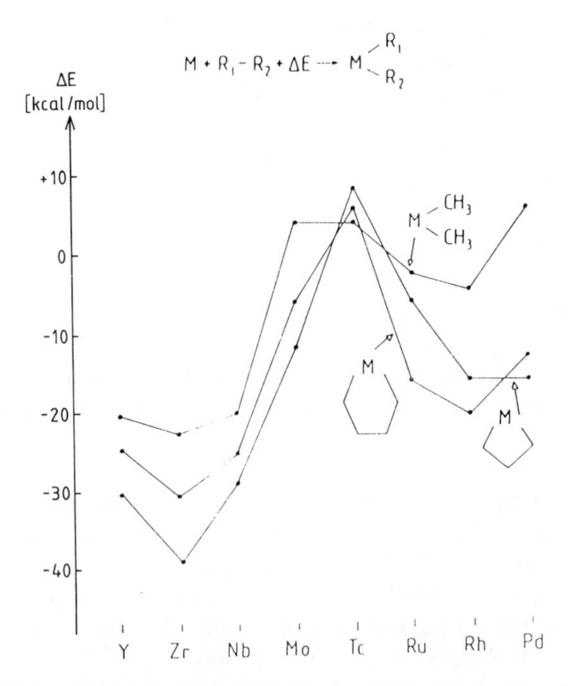

Fig. 8. Insertion product energies for the C-C activation of ethane, cyclopropane and cyclobutane, respectively. The energies are calculated relative to ground state reactants.

hydride complexes are thermodynamically favoured over the π-coordinated complexes. In fact, this may be true already for the zirconium atom for which the vinyl-hydride complex is favoured by 1–2 kcal/mol at the present level of accuracy. However, for the zirconium atom there is a computed barrier of 19 kcal/mol for breaking the C-H bond of ethylene, whereas π-coordination probably occurs without any barrier. In order not to have a too large kinetic advantage for forming the π-complex, ligands must be added to reduce the C-H activation barrier. This should be possible if covalently bonded ligands like hydrogen or chlorine atoms are added, since this will reduce the exchange loss in the reaction.

5. Activation of strained C-C bonds

The activation of unstrained C-C bonds through the oxidative addition mechanism has not yet been observed for any transition metal complex. As shown

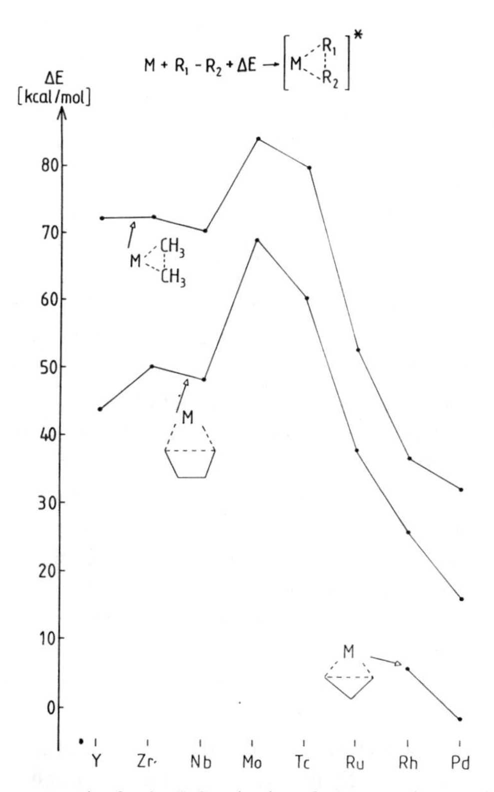

Fig. 9. Transition state energies for the C-C activation of ethane, cyclopropane and cyclobutane, respectively. The energies are calculated relative to ground state reactants.

above for the case of ethane, the barriers for breaking the C-C bond are very high for all second row transition metals due to the requirement for a simultaneous tilting of two methyl groups. However, activation of C-C bonds with strain has been observed in several cases, in particular, for cyclopropanes. In order to study the detailed differences of the activation of unstrained and strained C-C bonds, the reactions between second row transition metal atoms and ethane, cyclopropane and cyclobutane were compared in Ref. 30. The results of that study are given in Figure *8* for the reaction energies, and in Figure *9* for the barriers of the reactions.

When C-C bonds with strain are activated the reaction energies are, as expected, larger than when unstrained C-C bonds are broken. However, the reaction energy for the C-C bond breaking reaction is not as much larger for the strained compared to the unstrained case as the difference in C-C bond strength would indicate. The C-C binding energy of ethane was calculated to be 91.0 kcal/mol compared to two methyl radicals. The C-C bond strength of cyclopropane is only 60.9 kcal/mol, with respect to the C_3H_6 triplet diradical, and the C-C bond in cyclobutane is slightly stronger with 64.5 kcal/mol, with respect to the C_4H_8 triplet diradical. For the rightmost atoms the difference in the reaction energies between the ethane reaction on the one hand and the strained reactions for cyclopropane and cyclobutane on the other hand, is about 20 kcal/mol. For the atoms to the left this difference goes down to about 10 kcal/mol. The reason for the smaller difference between the strained and unstrained reaction energies than between the initial C-C bond strengths, is that there is strain involved also in the metallacycles that are formed after the strained C-C bonds have been broken. This strain is largest for the atoms to the left of the periodic table, since for these atoms the unstrained C-M-C angles are larger (115–180°) than they are for the atoms to the right. The strain in the metallacycles causes the C-M-C angle to be close to 90° for the metallacyclopentanes and about 70° for the metallacyclobutanes. The origin of the larger unstrained C-M-C angles to the left is that there is a large contribution of sp-mixing in the bonding for these atoms. For the atoms to the right, the bonds are formed with more pure sd-hybrids. Optimal sp-hybridized bonds have C-M-C angles of 180° whereas optimal sd-hybridized angles are 90°.

Of the hydrocarbons discussed here, ethane, cyclopropane and cyclobutane, the lowest barriers for breaking the C-C bond are found for cyclopropane. This is in line with the experimental observations of C-C bond activations by transition metal complexes [19]. The lower barrier for cyclopropane compared to cyclobutane is not due to a much weaker C-C bond strength for cyclopropane than for cyclobutane, see above. To understand this it is better to focus on the reverse reaction of elimination. The small activation barrier for the elimination reaction of cyclopropane is easy to understand from the highly strained metal-carbon bonds in the metallacyclobutane. For the metallacyclopentanes the equilibrium C-M-C angles are closer to the optimal bond angle, which explains the higher elimination barriers for these systems. Since the reaction energies for the cyclopropane and cyclobutane reactions are so similar, which is also due to differences in the metallacycle strain, a smaller elimination barrier as for cyclopropane will also lead to a smaller barrier for the C-C addition reaction.

The metal atom which has the lowest barrier for breaking the C-C bonds in all three systems studied here is palladium. This is connected with the fact that palladium is the only metal atom with an s^0 ground state, which is the state with the least repulsion towards ligands. The importance of the s^0 state was noted already for the breaking of the C-H bond in the previous section, where it was concluded that also a low lying s^1 state is necessary for a low activation barrier. The s^1 state is important for the final bond strength in the products. The same is true also for the activation of C-C bonds but in this case the s^0 state is even more important due to the larger repulsion towards the carbon centers. The slightly different relative importance of the s^0 and the s^1 states has the effect that rhodium has the lowest barriers for C-H activation reactions whereas palladium has the lowest barriers for C-C activation reactions.

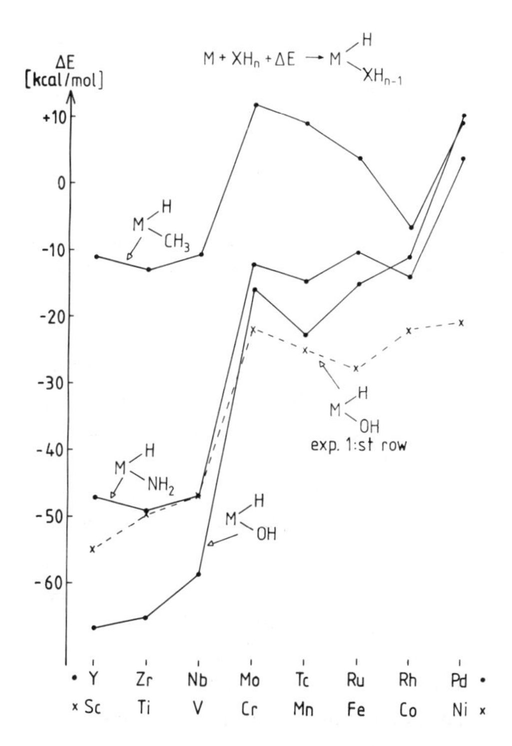

Fig. 10. Insertion product energies for the C-H, N-H and O-H activation of methane, ammonia and water, respectively. The energies are calculated relative to ground state reactants. Corresponding experimental values for the water insertion products of the first row transition metal atoms are also given [70].

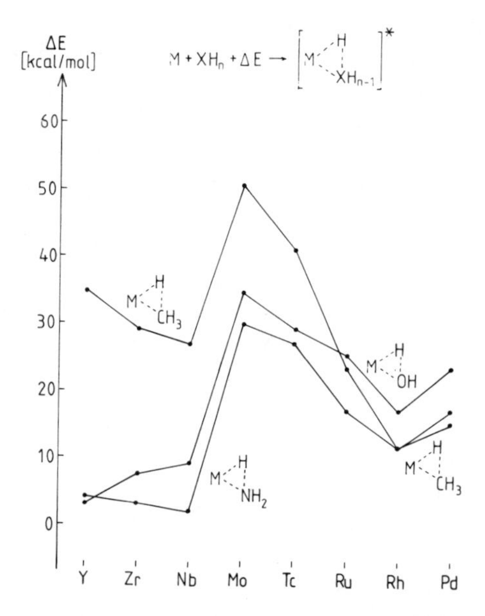

Fig. 11. Transition state energies for the C-H, N-H and O-H activation of methane, ammonia and water, respectively. The energies are calculated relative to ground state reactants.

It should finally be added that the calculations done to obtain the results in Figures *8–9* were done under the assumption of C_{2v} symmetry. This condition has since then been further investigated by releasing the symmetry constraint for the cyclobutane reactions. It then turns out that both equilibria and saddle points obtain the lower C_2 symmetry. The energy gains are 3–6 kcal/mol for all the metals and since this effect is so constant across the row, the trends in the figures are hardly modified in a noticeable way. The origin of the distortion from C_{2v} to C_2 is a repulsion at the C_b centra of the same type as the one that makes ethane staggered and not eclipsed. Another assumption made for most metals in Ref. 30 was that the electronic state at the saddle point is the same as at the equilibrium, where the ground state was carefully determined. This assumption turns out to be correct for most of the metals apart from yttrium. This system is still under investigation and for this case there appears to be a change of electronic state that might require a multiconfigurational treatment in the geometry optimization of the transition state. As discussed in the computational section, this situation is very rare, almost unique so far.

6. Lone-pair effects in oxidative addition reactions

The preceding sections have been devoted to reactions where H-H bonds, C-H bonds or C-C bonds have been activated. However, the oxidative addition of transition metal complexes to other types of bonds are also of practical and fundamental interest. From a fundamental point of view it is, for example, of interest to study the activation of bonds which are more ionic than C-H and C-C bonds and also interactions with systems where lone pairs are present. In an effort in this direction, the interaction between the entire sequence of second row transition metal atoms and the N-H bond of ammonia and the O-H bond in water were studied in Ref. 33 and 34, and the main results of those studies will be described here, see Figures *10* and *11*. Comparisons will be made to the methane reaction where no lone-pairs are present.

One major difference between the methane reaction and the water and ammonia reactions is that in the latter case substantially stronger bound molecular adducts will be formed which can act as precursors for the oxidative addition. For methane all complexes of the atoms except the palladium complex were found to be unbound. For water there is a slowly increasing trend of the binding energies going to the left, from 5 kcal/mol for palladium to 8 kcal/mol for zirconium. The complex for yttrium is much stronger bound with 15 kcal/mol. For ammonia the complexes both to the left and to the right in the row are found to be significantly more bound than the corresponding water complexes. The ammonia and water complexes in the middle of the row have rather similar binding energies.

It is interesting to note that even though methane lacks lone-pairs, the palladium atom is able to form a complex bound by 4 kcal/mol with methane. The binding in this system is due to an electron-nuclear attraction between a partially unshielded metal nucleus and the electrons on methane. The unshielding of the metal nucleus is an *sd*-hybridization effect due to the mixing between the s^0- and s^1-states of the metal, which moves metal electrons out into regions perpendicular to the metal-ligand axis. For the palladium-methane complex the s^0-state is the dominating state, which allows the metal atom to move as close as possible to methane, since this is the metal state which is the least repulsive towards ligands. The requirement to reduce this repulsion is the reason the only bound methane complex is found for palladium, since it is the only second row atom with an s^0 ground state. This type of bonding between an unshielded metal nucleus and the ligand electrons, dominate the bonding also for ammonia and water for the atoms to the right. For these systems this bonding is much more efficient since the lone-pair electrons of the ligands are easier to access for the metal than the methane bonding electrons. For this reason there is no longer an absolute requirement for an s^0 ground state and rhodium therefore forms an equally strongly bound complex with ammonia as palladium does.

The differences between ammonia and water are more intricate to understand than to understand the differences between these systems and methane. However, the differences can be understood from a combination of the facts that water has two lone-pairs but ammonia only one, and that the water lone pairs are more tightly bound, closer to the nucleus, than the one in ammonia. The more diffuse lone-pair of ammonia is the origin of the larger binding energies for these complexes than the ones for water. To the left, where the lone-pair donation bonding dominates, the zirconium and niobium complexes with ammonia have a binding energy almost three times that of the water complexes even though two water lone pairs can donate electrons. The yttrium atom is different and in this case the ammonia and water complexes have rather similar binding energies. For yttrium, the advantage of having two donating lone pairs, as in water, is maximized since the number of empty $4d$-orbitals is largest for yttrium.

The presence of the ammonia and water lone-pairs has a characteristic and large influence on the oxidative addition reaction. First, because of the lone-

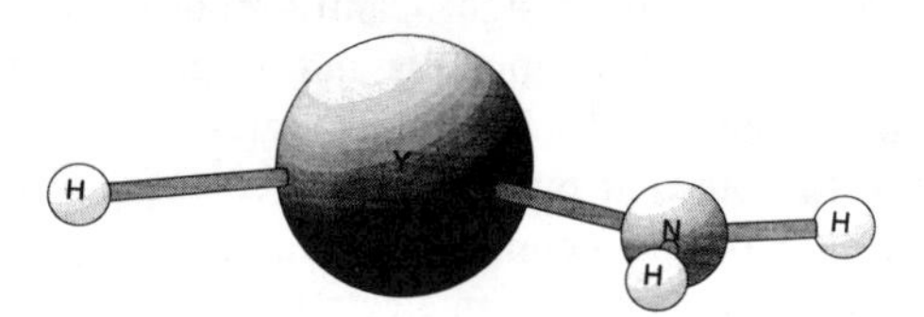

Fig. 12. Structure of $YHNH_2$.

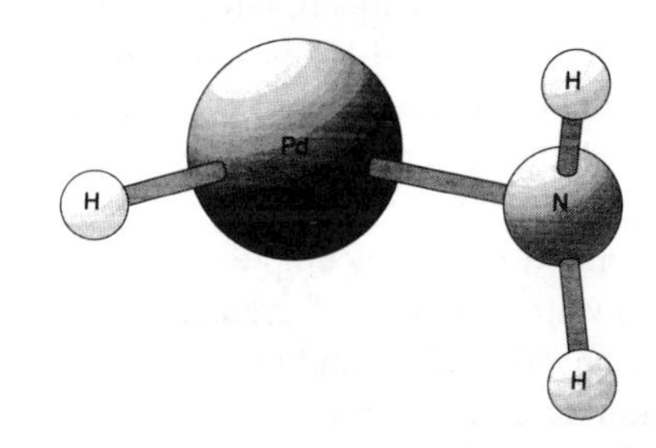

Fig. 13. Structure of $PdHNH_2$.

pairs, the geometries of the insertion products are quite different for the atoms to the left and for the atoms to the right in the periodic table. The insertion products for the atoms to the left in the row have approximately linear M-O-H and planar M-NH_2 subunits, respectively, while to the right the M-O-H subunit is bent and the M-NH_2 subunit is pyramidal, see Figures *12* and *13* where the ammonia insertion products are shown for a metal to the left and one to the right. To the right the geometries are such that the lone-pair points away from the metal to reduce the mainly repulsive interaction between the metal and the lone-pairs. To the left the interaction between the metal and the lone-pairs is attractive and in the equilibrium geometries the overlap between the lone-pairs and the empty metal 4*d* orbitals is maximized for these products.

The second characteristic effect of the lone pairs concern the trend of the energetics across the row, see Figure *10*. The lone-pairs have largest effect on the reactions for the metal atoms to the left in the row where there are empty 4*d*-orbitals. The dramatic influence of empty, rather than half-empty, 4*d*-orbitals is best seen on the binding energies of the insertion products. There is a very large increase in binding energy going from the molybdenum to the niobium products for both ammonia and water, of 35 kcal/mol and 43 kcal/mol, respectively. This increase occurs at exactly the same point as the onset of empty 4*d*-orbitals. For the product energies to the left of the row, where the interaction with the lone-pairs is mainly attractive, the water product binding energies are larger than the ones for ammonia, due to the presence of two lone pairs for water. Since the bond distance is shorter for the insertion products than for the molecular adducts, discussed above, the fact that the lone-pairs of water are more compact does not matter as much for the interaction energy of the insertion products. To the right of the row where the repulsive interaction between the metal and the lone-pairs dominates, the insertion products energies are rather similar for the two reactions, which indicates that part of the repulsion is avoided in the final geometry.

For the C-H activation of methane the lowest barrier was obtained for the rhodium atom. This is due to an optimal mixing of low-lying s^1-and s^0-states for this atom, which allows both for strong bond formation and close approach that makes the interaction with the C-H bond very effective. Of the atoms to the right, rhodium continues to be the one that easiest breaks the bonds for both the ammonia and the water reaction, see Figure *11*. This is due to a similar bond activation mechanism for all three reactions of a basically covalent type where the lone-pairs of the ligands have little influence. The situation is quite different for the atoms to the left. For these atoms, the lone-pairs assist in a marked way in the bond-breaking. Since the ligands are further out from the metal for the transition states than for the products, the fact that the ammonia lone-pair is diffuse also is of definite importance. This leads to a lower barrier for the ammonia reaction than for the water reaction. The barriers for the two reactions are, however, very similar for the case of yttrium, where the repulsive effects between the lone-pairs and the metal is smallest. For this atom the interaction with two lone-pairs, even though they are more tightly bound, rather than with one is an advantage and the barrier for the water reaction is therefore very close to the one for the ammonia reaction. The calculations in Ref. 34 indicate, in fact, that the barrier for breaking the O-H bond in water should be close to zero in this case.

Most of the present review is concerned with results obtained for the second row transition metals. However, for the oxidative addition reaction of water a few studies exist which in an illustrative way shows a general and interesting difference between the first and second row transition metals. Margrave and coworkers [70] have studied the O-H insertion reaction of water by first row transition metal atoms using matrix isolation spectroscopy. Measured IR frequencies lead to estimates of the insertion product binding energies shown in Figure *10* together with the theoretical results for the second row transition metals discussed above. As can be seen from this figure, the insertion products for the first row metals are less bound to the left and more bound to the right compared to those of the second row. This can be rationalized in terms of a dominating interaction between the lone-pairs and the metal d-orbitals, combined with the fact that the d-orbitals are much smaller than the valence s-orbitals for the first row but more similar in size for the second row. Since the interaction to the left of the row is attractive, it is an advantage with a close interaction between the lone-pairs and the metal d-orbitals to the left and this explains the larger binding energies for the second row metals in this region. The situation to the right is the opposite with a dominatingly repulsive interaction and therefore the binding energies are larger for the first row atoms in this region. The much larger repulsive interaction between the water lone-pairs and the metal d-orbitals for the second row than for the first row is also demonstrated in a combined theoretical and experimental study of the O-H insertion reaction between a nickel atom and water [60]. This reaction was shown to proceed without any barrier. In contrast, the reaction between palladium and water has a substantial barrier of more than 20 kcal/mol. These results also show the effectiveness of sd-hybridization, since the s^0-state is much higher in energy for nickel than for palladium and still the barrier is lower for nickel. The nickel atom is able to approach water with a low repulsion by sd-hybridizing away the repulsive 4*s*-electron.

7. Methane activation by transition metal cations

Experiment on naked metal atoms indicate that cations are more reactive towards alkanes than neutrals. By comparing the methane activation by naked transition metal cations with the same process for the naked neutrals further understanding of the oxidative addition reaction mechanisms can be obtained. Also, the large amount of experimental information available on the gas phase reactions between naked metal cations and alkanes gives an excellent opportunity to calibrate the results obtained from the calculations. In these experimentally observed reactions neutral fragments of the alkanes,

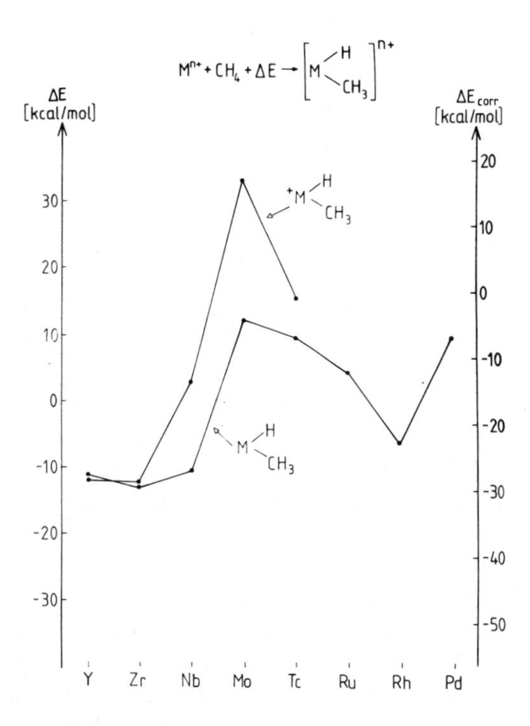

Fig. 14. Insertion product energies for the C-H activation of methane by neutral and cationic metal atoms, respectively. The energies are calculated relative to ground state reactants. The DE_{corr} values include the estimated bond energy corrections given in Section 2.3 and corrections for zero point vibrational energy differences.

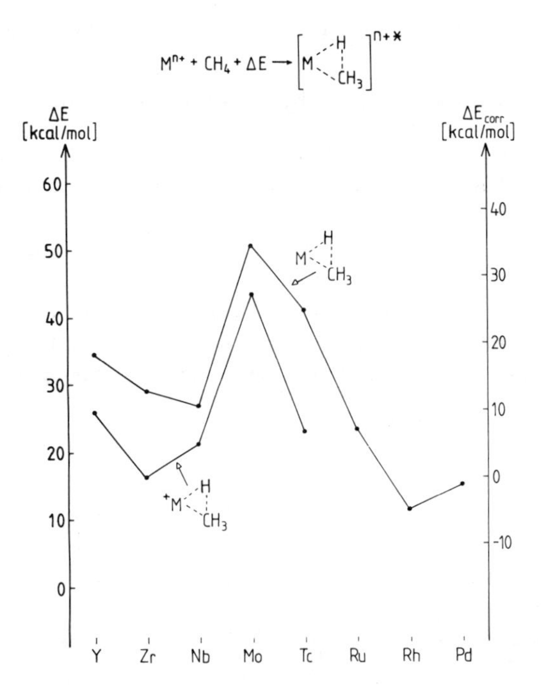

Fig. 15. Transition state energies for the C-H activation of methane by neutral and cationic metal atoms, respectively. The energies are calculated relative to ground state reactants. For the DE_{corr} values see Fig. *14*.

e.g. H_2, are normally eliminated and only the final cationic elimination products are observed. However, the experimental results indicate that the first step in these reactions is the oxidative addition of the cation into a C-H bond and the barrier for this step is thus an important part of the observed reactions. Therefore the methane activation by naked second row metal cations have been studied [66] and the results are shown in Figures *14* and *15*.

As discussed above in Section 3 there are two factors determining the barrier heights for the neutral atoms in their reactions with methane. First, there is the repulsion between the metal electrons and the closed shell electrons of methane, which is minimized for the metal s^0 states. Secondly, the covalent bonding in the final insertion product has to build up in the transition state region, and this is optimal for the s^1 state of the metal atoms. The most obvious difference between the cationic and the neutral metal atoms in their interaction with methane is the existence of a long-range electrostatic attraction between the cations and the polarizable alkane, leading to the formation of molecularly bound complexes for all the second row cations, bound by up to 15 kcal/mol relative to the ground state atomic dissociation limit. For the neutral second row metals a weakly bound such a complex is formed only for the palladium atom. It has often been assumed that this attraction between the cation and the alkane can be considered as an extra attractive factor in the transition state region, shifting down an otherwise constant barrier for the corresponding reaction of the neutral atom by the interaction energy of the molecular complex. Comparing the barrier heights for the C-H insertion reaction between the cations and the neutrals it can be found that for yttrium to technetium the barrier is actually lower for the cations than for the neutrals, see Figure *15*, even though the energy difference in most of the cases do not correspond to the binding energy of the molecular complex. However, for rhodium the barrier is substantially higher for the cation than for the neutral and for ruthenium and palladium the potential surfaces for the cations are completely different from those of the neutrals and no stable insertion products exist. Furthermore, for all second row metals except yttrium the insertion product is less bound for the cations than for the neutrals, see Figure *14*. This means that the differences between the cations and the neutrals in their reaction mechanism for C-H insertion is not only a charge effect, but also the differences in atomic spectra between the neutral and the cation has to be taken into account.

The decreased bonding in the insertion product for most second row cations compared to the neutrals can be explained by the lower *s*-populations in the cationic ground states. The s^0 state binds much less efficiently than the s^1 state and for ruthenium and palladium, having s^0 ground states, there are not even minima in the region of the insertion product structure. The decreased repulsion in the transition state region for the cations is, apart from the attractive charge effect discussed above, also caused by the lower *s*-population in the atomic ground state of the cations. However, for cations like niobium and molybdenum the decrease in repulsion on going from an s^1 ground state for the neutral atom to an s^0 ground state for the cation is to a large extent counterbalanced by a decrease in built up covalent bonding, due to the lower bonding capacity of the s^0 state. The balance between the bonding capacity and the amount of repulsion for different atomic states has a general effect on the shape of the potential surfaces which can most easily be seen by considering the reverse reaction, the reductive elimination of methane. For this reaction there is a trend towards lower elimination barriers for the cations compared to the neutral metal atoms. Since there is also a trend towards lower elimination barriers going from the left to the right in the periodic table, from 46 kcal/mol for the yttrium atom to 6 kcal/mol for the palladium atom, the elimination barrier for the atoms to the right is actually eliminated by going to the cation. Thus the low bonding capacity of the s^0 state together with high promotion energies to the s^1 state leads to a different shape of the potential curves for the metals to the right. For the ruthenium and palladium cations there is no elimination barrier and consequently no insertion product, the only minima on these potential surfaces are those for the molecular complexes. For rhodium a very unstable insertion product was found, having a low elimination barrier and a transition state structure very different from the rest of the atoms.

All second row metal cations studied experimentally have been found to be unreactive with methane, while all these metals do react with larger alkanes. The most favorable pathway for the reaction of metal cations with methane is: $M^+ + CH_4 \rightarrow MCH_2^+ + H_2$. This reaction is expected to be endothermic for most metals, which is assumed to be the main reason methane activation by metal cations has not been observed. In order to test the accuracy of our calculations the endothermicity of this reaction was calculated for one particular case, yttrium, where explicit experimental results are available. If the estimated bond energy errors from Section 2.3 are applied and corrections for zero point vibrational differences are taken into account, the elimination of H_2 from methane by the yttrium cation is calculated to be endothermic by 19 kcal/mol. The experimental value is 15 ± 3 kcal/mol [71], and a reasonable agreement between theory and experiment is thus obtained. Furthermore, the C-H insertion barrier is in the same way estimated to be 10 kcal/mol for the yttrium cation, E_{corr} in Figure *15*, which is thus lower than the endothermicity for the elimination reaction, supporting the validity of the determination of the Y^+-CH_2 binding energy from the experimental reaction endothermicity.

8. Ligand effects on the oxidative addition reaction

So far most of the general trends studied for the oxidative addition reaction have been best illustrated by results obtained for ligand-free metals. The present type of study, is not limited to naked atoms and in this section some of the more recent results for the effects of ligands on the oxidative addition will be described. Ligand effects on the reactivity of transition metal complexes can be divided into essentially two different classes. Ligand effects that are of electronic origin belong to the first class and those which have a

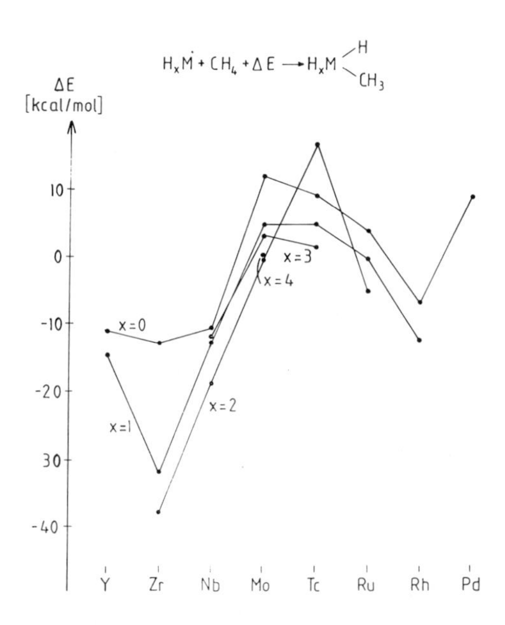

Fig. 16. Insertion product energies for the C-H activation of methane by MH_x, $x = 0$–4. The energies are calculated relative to ground state reactants.

steric origin belong to the second class. The optimal theoretical treatment of these two types of effects are quite different. The electronic structure effects from ligands require highly accurate theoretical methods, including a treatment of correlation effects of all valence electrons using large basis sets. This type of methods are not required when ligand effects of basically steric origin are treated. In fact, if the steric effect is just a blocking of a coordination site, no calculations at all are needed but a simple reasoning about the effects is often sufficient. In more complicated situations, steric effects are best handled by classical methods like molecular mechanics [72]. At the end, the results of these different treatments have to be combined into a unified picture of ligand effects. In the present review, the discussion will focus on the electronic structure effects of ligands.

As a first step in the study of the electronic structure aspects of ligands, the effects on the methane reaction from adding covalently bound ligands on the metal will be discussed. Covalent ligands here mean ligands that change the oxidation state of the metal. For each metal atom the covalency is systematically

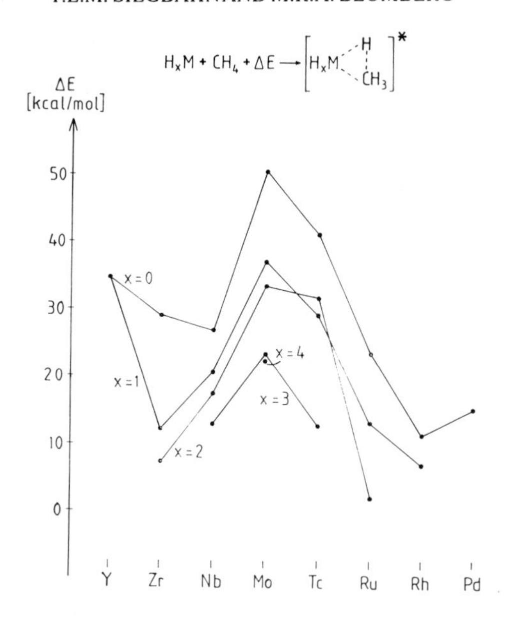

Fig. 17. Transition state energies for the C-H activation of methane by MH_x, $x = 0$–4. The energies are calculated relative to ground state reactants.

saturated by adding hydrogen ligands one by one to the complex. As a parallel to the methane reaction, the results for the H_2 addition reaction for the same metal complexes are also discussed. The main motivation for choosing covalently bound ligands for the present study is connected with the exchange energy loss discussed in the previous sections above. For every covalently bound ligand added to the metal, the exchange energy loss in the oxidative addition reaction will decrease. If this was the only effect from adding these ligands, one should expect both lower addition barriers and more strongly bound products the more ligands that are added. It should be added that the important point is that the spin state of the metal is reduced by the addition of ligands, and the simplest way to achieve this is to add covalently bound hydrogen ligands. The lowering of the spin can also occur when lone pair ligands are added. For example, when CO is added to the nickel atom the spin is reduced from triplet to singlet. However, the addition of lone pair ligands will not lead to such a systematic lowering of the spin as when covalent ligands are added.

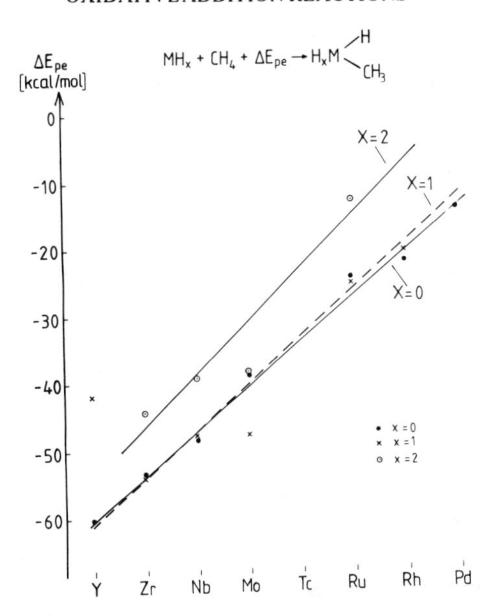

Fig. 18. Insertion product energies for the C-H activation of methane by MH_x, $x = 0$–2. Promotion and exchange energies are subtracted from the reaction energies.

The results for the ligand effects on the oxidative addition of methane are shown in Figures *16* and *17*. As is evident from these figures, the general shapes of the curves are the same as in the ligand free case, with a minimum in binding energies and maximum in barrier heights in the middle of the row. This means that loss of exchange and promotion effects still dominate the general behaviour. In order to see more clearly other effects on the oxidative addition reaction generated by the ligands, the promotion and exchange loss energies are subtracted from the results and the new results are shown in Figure *18*. The estimates of the exchange energy loss energies were obtained from tables of exchange energies given by Carter and Goddard [45]. If exchange and promotion effects had been the only effects of importance the curves in Figure *18* would have been straight constant lines. Since this is far from the case, it means that other effects are also of large importance. In Figure *18* two new trends can be identified. First, there is a systematic decrease of the reaction energies as one goes from left to right in the periodic table. This trend is explained by the dominant role played by electron repulsion between the metal electrons and the ligand electrons, which increases with the increasing number of metal electrons to the right in the row. The second trend is a systematic decrease of the reaction energies as the number of hydrogen ligands is increased. This trend is intuitively explained by simply the direct steric repulsion effects between the ligands. However, since this effect is practically identical for the H_2 and the CH_4 reaction it appears that the effect is dominated by local rehybridization on the metal, which should be the same for hydrogen and methyl ligands.

The comparison between the H_2 and CH_4 reactions provides additional insight into the dominating energetic effects in the oxidative addition reaction. The difference between the reaction energies for these reactions increase to the right in the periodic table (see Figure *1*). This trend is again best explained by the important role of the electronic repulsion effect between the metal electrons and the electrons on the ligands. Methyl has more electrons than hydrogen and this repulsion is therefore larger for methyl and larger to the right in the periodic table, which leads to an increased difference between the reaction energies of H_2 and CH_4 to the right. The second trend in the difference between the reaction energies of H_2 and CH_4 is more surprising. Even though methyl is bulkier than the hydrogen atom, the difference in reaction energy between H_2 and CH_4 decreases as the number of ligands increases. This counterintuitive trend is explained by the electronegative character of the hydrogen and methyl ligands. This means that the more ligands that are added the more electrons are moved from the metal to the ligands, which in turn means that the direct repulsive effect between the electrons on the metal and the electrons on the ligands should decrease. Since this repulsive effect is the dominating origin of the difference in the reaction energies between H_2 and CH_4, this difference will also decrease as more ligands are added.

The lowest barriers obtained for the oxidative addition of methane are of particular interest in comparison to what is known experimentally. The second lowest barrier of all reactions studied here is obtained for RhH, which is a Rh(I) complex. This is in line with the fact that the only second row transition metal complexes which are found to dissociate the C-H bond in alkanes are Rh(I) complexes. However, the lowest barrier for this reaction obtained in the present study actually occurs for RuH_2. In fact, when corrections for a larger basis set and configuration expansion are made, RuH_2 is predicted not to have any barrier for this reaction.

In a recent study, halide ligand effects have been studied by replacing the hydride ligands in the systems described above by halide ligands [73]. The particular effects of halide ligands are best seen by focussing on the differences

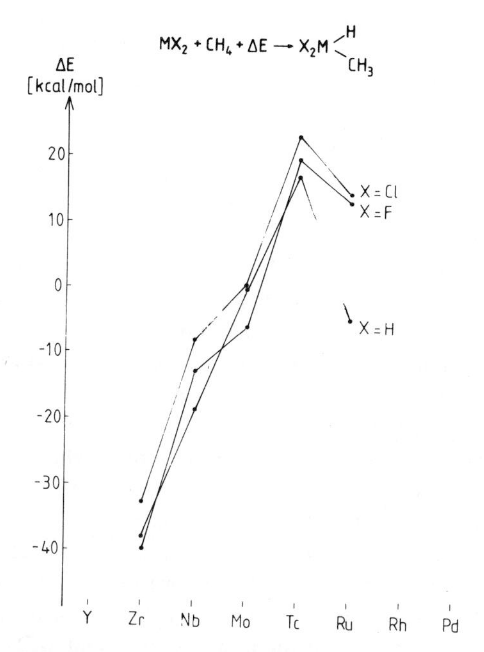

Fig. 19. Insertion product energies for the C-H activation of methane by MX_2, X = H,F,Cl. The energies are calculated relative to ground state reactants.

of the results of having hydride and halide ligands. For example, in this way the common effects of promotion and exchange will to a large extent be cancelled for most systems. The most interesting differences between halide and hydride ligands are that halide ligands are more electronegative and that they also have electron donating capability. Some results for the halide ligand effects on the methane reaction are given in Figure *19*.

A few main conclusions can be drawn from the halide results. First, with a few exceptions the results for the halide complexes are very similar to those for the hydride complexes. This is true, in particular, for the complexes of the atoms to the left where the bonding is more ionic. Secondly, there is a notable trend towards a destabilizing effect of halide ligands to the right in the periodic table. The origin of this effect is a more covalent bonding to the right with the 5*s*-electrons playing a key role. The 5*s*-electrons have the lowest orbital energies and are therefore the ones that are preferentially donated to the halide ligands, but they are also needed to form proper *sd*-hybrids to bind the R-groups. The consequence of this competition for the metal 5*s* electrons can be noticed already for the simplest possible hydrides and halides. For example, for RhH_2 the first and the second hydrogen bind about equally strongly with energies of 64.1 kcal/mol [35] and 65.2 kcal/mol [37]. In contrast, for RhHCl the hydrogen is bound by only 39.7 kcal/mol [73]. Another way to understand these results is to consider the metal as a cation when halide ligands are present. Since the bonding s^1-state is higher in energy for the cations than for the neutral atoms, the cations will form weaker bonds than the neutral atoms (see also Section 7 for the cations).

Halides bind with more ionic bonds than the hydrides and this has two immediate simple consequences of importance for the oxidative addition reaction. First, when a 4*d*-electron is donated to a ligand this will lead to an increase of the spin for the atoms to the right where the 4*d*-shell is more than half-filled. For the atoms to the left there will instead be a decrease of the spin when a 4*d*-electron is donated. In contrast, when a covalent bond is formed there will be a decrease of the spin both to the left and to the right. Transition metal halides will therefore in general have the same spin as the hydrides for the atoms to the left, but there is a tendency towards higher spin for the halides compared to the hydrides for the atoms to the right, since the hydrides are more covalently bound. In other words, exchanging hydrides with halides for complexes of the atoms to the right will bring down high-spin states in comparison to low-spin states. Since one of the conditions for a low barrier for the oxidative addition reaction is that there are two low-lying states of different spin, this will be an advantage for the halide complexes in cases where the high-spin state of the hydrides is high in energy. This is, for example, the case for the $RhCl(PH_3)_2$ complex, which has recently been studied. In the opposite situation, where the low-spin state is high in energy for the hydride, it will be a disadvantage for the oxidative addition to replace the hydrides with halides. The second important effect of the more ionic bonds for the halides is that the metal 5*s*,5*p*-population will tend to be smaller than for the hydrides. This is a disadvantage, as already mentioned above, for the high-spin state since this state should bind the products and there will be a competition for these bonding electrons between the halides and the R-groups of the dissociated reactant. For the low-spin state the low metal 5*s*,5*p*-population for the halides is an advantage, since this allows the metal to approach the reactant more closely and thus interact more effectively with the bond to be broken.

In the above analysis of ligand effects the important class of lone pair ligands have not yet been discussed. A few preliminary results available for the combined effects of hydride, halide and lone pair ligands will therefore finally be mentioned here. Only the oxidative addition of H_2 has been studied so far. If the H_2 reaction energies for the reactions with MH(CO) and MCl(CO) are compared, two main conclusions can be drawn. First, for the atoms to the left the reaction energies of these two reactions are quite similar. This is the same conclusion as drawn above for the case without lone pair ligands, namely that for the complexes to the left hydride and halide ligands have roughly the same effect on the oxidative addition reaction. However, for the atoms to the right the presence of lone pair ligands reverses the effects. Without lone pair ligands, halides have a *destabilizing* effect on the reaction energies as discussed above, but with lone-pair ligands present, halide ligands have a *stabilizing* effect on the reaction. For rhodium, the effect of exchanging a hydride ligand with a chloride ligand is +14.1 kcal/mol on the reaction energy without the carbonyl ligand present, which changes to –10.2 kcal/mol when carbonyl is present. The origin of this effect is best explained by focussing on the binding of the carbonyl. For systems like MH(CO) the binding of carbonyl has large covalent components through donation backdonation type bonding. When this system reacts with H_2, the metal electrons in the product $MH_3(CO)$ are covalently tied up in the bonding to the hydrogens and the covalent bonding to carbonyl is therefore partly destroyed. This leads to a large *loss* of metal-carbonyl binding energy in the reaction. On the other hand, for systems like MCl(CO) the metal-carbonyl binding is largely ionic. When H_2 is added to MCl(CO) the ionic bonding of the carbonyl is even more efficient since the added hydrogens also remove electrons from the metal and thereby increase the positive charge of the metal. This leads to a substantial *gain* in metal-carbonyl binding energy in the reaction. For the atoms to the left there is almost no difference between the MH(CO) and MCl(CO) systems since for these systems the metal-carbonyl bonding is about equally ionic. Also, the ionic bonding is saturated already with one ligand and therefore does not increase as it does to the right when H_2 is added. The reason for this difference between the atoms to the left and to the right is that the positive charge of the core (including the $1s$–$4p$ electrons) is much smaller for the atoms to the left.

9. Activation of other systems

In this section the results for a few other oxidative addition reactions, which have also been studied, will be briefly discussed. The oxidative addition of an Si-H bond to a transition metal center is known for a number of different transition metals [1] and is believed to be the key activation step in the important hydrosilation reactions. Several metals are known to homogeneously catalyze the hydrosilation of olefins, acetylenes and ketones and it has been observed that the oxidative addition of an Si-H bond to a metal center is similar to the addition of H_2 [1, 74], i.e. the reaction occurs readily and stable insertion products are formed. The Si-H bond in SiH_4 is calculated to be 17 kcal/mol weaker than the C-H bond of CH_4 and about 15 kcal/mol weaker than the H-H bond of H_2. The activation of an Si-H bond could therefore be expected to be easier than that of an alkane C-H bond. To investigate if there is such a trend the oxidative addition of silane to all second row transition metal atoms was investigated [75].

The silane activation is found to be more exothermic than the methane activation for all second row metals, and the difference is larger for the metals to the right in the periodic table. The silane and the H_2 activations are found to have almost identical reaction energies for the metals yttrium to technetium, which means that the M-Si bond for these metals is close to 15 kcal/mol weaker than the M-H bond. For ruthenium to palladium the silane activation is more exothermic than the H_2 activation by 7–12 kcal/mol. Thus, for palladium the M-Si bond is almost as strong as the M-H bond. Compared to the metal-methyl bond the metal-silyl bond is weaker for the metals yttrium to technetium and stronger for ruthenium to palladium.

As discussed in the previous section, if promotion and exchange energies are taken into account, there is a general trend for MR_1R_2 binding energies to become smaller going from the left to the right in the periodic table, see Figure *18*. One explanation for this trend is the increased repulsion between the ligand electrons and the metal valence electrons to the right, due to the larger number of *d*-electrons. Another factor that might contribute to this trend is the decreased net metal-to-ligand charge donation going to the right, due to the larger ionization energies to the right. Comparing the $MHSiH_3$ binding energies to both the $MHCH_3$ and the MH_2 binding energies the decrease to the right is much smaller for $MHSiH_3$. The most likely explanation for this difference is that the repulsion to the right does not increase as much for the metal-silyl bond, first because the M-Si bond is longer than both the M-H and the M-C bond, and secondly because the SiH bonding electrons are more displaced towards the hydrogens than in the corresponding CH bonds, and thus further away from the repulsive metal electrons. Concerning the net metal to ligand donation it is slightly smaller for the silane case compared to the methane case for all metals (by about 0.1 electron), but there is no difference between the systems to the left and to the right in this respect. Therefore, electron donation from the metal does not appear to explain the large metal-silyl binding energies to the right.

The activation barrier for metal insertion into the Si-H bond is for all metals found to be lower than for the corresponding methane reaction. In particular for the metals to the right, ruthenium to palladium, there seems to be no barrier for the Si-H insertion reaction. For all the other metals except molybdenum the calculated barrier for silane activation is low, indicating that this reaction will proceed rather easily for most of the second row metals.

During the course of this project [28–42] several other oxidative addition reactions have been studied. A few of these will be very briefly mentioned here. An interesting class of reactions is the one where double bonds are activated. This is, for example, the case for the oxidative addition of O_2. Due to the very large metal-oxygen bond strengths to the left this reaction is quite exothermic for these metal atoms. The reaction proceeds through molecularly bound peroxo- or superoxo-complexes [42]. The activation of the C-O bond in CO_2 is another reaction which is very exothermic for the metals to the left. In fact, the reaction between zirconium and CO_2 is one of a few reactions studied so far, where the geometry optimization at the SCF level went downhill in energy from the starting point as a molecular complex all the way to the dissociated insertion product. The activation of the oxygen-carbon bonds in ethylene-oxide proceeds through a four membered metallacycle and then over an activation barrier, which is low for the metals to the left, to a metal-oxide with a π-bonded olefin. This reaction is exothermic for all second row transition metal atoms except for palladium, which means that the reverse reaction of olefin epoxidation will not occur for these metal atoms. Other oxidative addition reactions where results are available are the activation of the halide-halide bonds in F_2 and Cl_2, and the activation of HCl and CH_3Cl. All these reactions are strongly exothermic and proceed without any barriers. Finally a reaction will be mentioned that is normally not thought of as an oxidative addition reaction and this is the π-coordination of olefins. For the metal atoms to the left the C-C bond changes from a double bond to a typical single bond in this reaction, which means that it can be considered to be an oxidative addition reaction of the C-C π-bond to the transition metal [29].

10. Summary

In the above sections the oxidative addition reaction has been reviewed. It has been shown that the alkane C-C bond is more difficult to activate than the alkane C-H bond, which in turn is more difficult to activate than the H-H bond. This is due to the directional character of the methyl group in contrast to the spherical nature of the hydrogen atom. C-H activation becomes simpler for unsaturated hydrocarbons, partly because the geometrical hindrance for the C-H attack is less severe for these systems. Also, the π- and π^*-orbitals of the unsaturated hydrocarbons increase the exothermicity of the reactions which has a reducing effect on the barriers. Strained C-C bonds are as expected easier to activate than unstrained C-C bonds. The origin of the lower barrier for C-C activation of cyclopropane than for cyclobutane is a combination of a similar reaction energy for the two systems and a lower elimination barrier for cyclopropane. The elimination barrier is lower for cyclopropane because the C-M-C angle is much smaller in the product of this reaction. Lone-pairs have a substantial effect on the oxidative addition reaction as illustrated on the sequence of O-H, N-H and C-H bonds in water, ammonia and methane, respectively. The lone-pair effect is strongly attractive to the left and repulsive to the right. For the C-H activation of alkanes the metal atoms to the right have lower barriers than those to the left, due to the presence of the low-lying s^0-state for the atoms to the right. Rhodium is particularly efficient since for this atom both the s^0- and the s^1-states are low-lying. Covalent ligands tend to reduce the size of the barriers since exchange loss energies are reduced. When lone-pair ligands like carbonyl are present it is a big advantage for the atoms to the right to have halide ligands present or conversely when halide ligands are present it is a big advantage to have lone-pair ligands present. The reason for this is that with halide ligands present, the bonding to the lone-pair ligand is ionic and this ionic bonding is stabilized further for the atoms to the right when the reactant alkane is added. It is interesting to note that when the favourable properties for alkane C-H activation mentioned above are added together, a complex of the type RhCl(CO) appears, strikingly similar to the system $RhCl(PPh_3)_2$ which is one of the few systems that has been found to activate alkanes experimentally.

The present review has focussed on the study of trends starting with results for naked metal atoms. The second row of the transition metals has been chosen for the studies since this is the row that is easiest to treat using the present type of *ab initio* methods. Similar results are available also for other types of reactions such as carbonyl and olefin insertions. In the near future the same type of studies will be carried out also for the first transition metal row, possibly with the use of local density methods which appear to be very promising for the treatment of general transition metal complexes.

P. E. M. Siegbahn and M. R. A. Blomberg,
Department of Physics,
University of Stockholm, Box 6730, S-113 85 Stockholm, Sweden

References

1. J.P. Collman, L.S. Hegedus, J.R. Norton and R.G. Finke: Principles and Applications of Organotransition Metal Chemistry. University Science Books (1987), Mill Valley
2. J. Chatt and J.M. Davidson, J. Chem. Soc. 1965, 843
3. W.D. Jones and F.J. Feher, Acc. Chem. Res. 22, 91 (1989)
4. (a) A.H. Janowicz and R.G. Bergman, J. Am. Chem. Soc. 104, 352 (1982); (b) A.H. Janowicz and R.G. Bergman, J. Am. Chem. Soc. 105, 3929 (1983)
5. (a) J.K. Hoyano and W.A.G. Graham, J. Am. Chem. Soc. 104, 3723 (1982); (b) J.K. Hoyano, A.D. McMaster and W.A.G. Graham, J. Am. Chem. Soc. 105, 7190 (1983)
6. (a) W.D. Jones and F.J. Feher, J. Am. Chem. Soc. 104, 4240 (1982); (b) W.D. Jones and F.J. Feher, J. Am. Chem. Soc. 106, 1650 (1984)
7. T. Sakakura, T. Sodeyama, K. Sasaki, K. Wada and M. Tanaka, J. Am. Chem. Soc. 112, 7221 (1990)
8. Perspectives in the Selective Activation of C-H and C-C Bonds in Saturated Hydrocarbons, B. Meunier and B. Chaudret, Ed., Scientific Affairs Division – NATO (1988), Brussels
9. A.J. Rest, I. Whitwell, W.A.G. Graham, J.K. Hoyano and A.D.J. McMaster, Chem. Soc. Chem. Commun. **1984**, 624
10. D.M. Haddleton, A. McCamley and R.N. Perutz, J. Am. Chem. Soc. **110**, 1810 (1988)
11. S.T. Belt, F.-W. Grevels, W.E. Klotzbücher, A. McCamley and R.N. Perutz, J. Am. Chem. Soc. 111, 8373 (1989)
12. J. Halpern, Inorganica Chimica Acta **100**, 41 (1985)
13. P.O. Stoutland and R.G. Bergman, J. Am. Chem. Soc. **107**, 4581 (1985)
14. P.O. Stoutland and R.G. Bergman, J. Am. Chem. Soc. **110**, 5732 (1988)
15. (a) M.V. Baker and L.D. Field, J. Am. Chem. Soc. **108**, 7433 (1986); (b) M.V. Baker and L.D. Field, J. Am. Chem. Soc. **108**, 7436 (1986); (c) T.T. Wenzel and R.G. Bergman, J. Am. Chem. Soc. **108**, 4856 (1986)
16. W.D. Jones, R.M. Chin, L. Dong, S.B. Duckett and E.T. Hessell, In Energetics of Organometallic Species, Simões, J.A.M., Ed., Kluwer (1992), Dordrecht, pp. 54–67
17. C.F.H. Tipper, J. Chem. Soc. **1955**, 2045
18. D.M. Adams, J. Chatt, R. Guy and N. Sheppard, J. Chem. Soc. **1961**, 738
19. R.H. Crabtree, Chem. Rev. **85**, 245 (1985)
20. (a) Z.H. Kafafi, In Selective Hydrocarbon Activation: Principles and Progress, J.A. Davies, P.L. Watson, A. Greenberg and J.F. Liebman, Ed., VCH Publishers (1990), New York, pp. 411–432; (b) E.S. Kline, R.H. Hauge, Z.H. Kafafi and J.L. Margrave, Organometallics **7**, 1512 (1988)
21. (a) K.J. Klabunde and Y. Tanaka, J. Am. Chem. Soc. 105, 3544 (1983); (b) K.J. Klabunde, Gi Ho Jeong and A.W. Olsen, In Selective Hydrocarbon Activation: Principles and Progress, J.A. Davies, P.L. Watson, A. Greenberg and J.F. Liebman, Ed., VCH Publishers (1990), New York, pp. 433–466
22. D. Ritter and J.C. Weisshaar, J. Am. Chem. Soc. **112**, 6425 (1990)
23. S.A. Mitchell and P.A. Hackett, J. Chem. Phys. **93**, 7822 (1990)
24. (a) J.C. Weisshaar, Advances in Chem. Phys. 81; (b) J.C. Weisshaar, Ch.13 in Gas-Phase Metal Reactions, A. Fontijn, Ed., Elsevier (1992), Amsterdam
25. P.B. Armentrout and J.L. Beauchamp, Acc. Chem. Res. **22**, 315 (1989)
26. P.B. Armentrout, In Selective Hydrocarbon Activation: Principles and Progress, J.A. Davies,

P.L. Watson, A. Greenberg and J.F. Liebman, Ed., VCH Publishers (1990), New York, pp. 467–533

27. P.A.M. van Koppen, M.T. Bowers, J.L. Beachaump and D.V. Dearden, In Bonding Energetics in Organometallic Compounds, T.J. Marks, Ed., ACS Symposium Series (1990), Washington DC, pp. 34–54
28. M.R.A. Blomberg, P.E.M. Siegbahn and M. Svensson, J. Am. Chem. Soc. **114**, 6095 (1992)
29. M.R.A. Blomberg, P.E.M. Siegbahn and M. Svensson, J. Phys. Chem. **96**, 9794 (1992)
30. P.E.M. Siegbahn and M.R.A. Blomberg, J. Am. Chem. Soc. **114**, 10548 (1992)
31. P.E.M. Siegbahn, M.R.A. Blomberg and M. Svensson, J. Am. Chem. Soc. **115**, 1952 (1993)
32. P.E.M. Siegbahn, M.R.A. Blomberg and M. Svensson, J. Am. Chem. Soc. **115**, 4191 (1993)
33. M.R.A. Blomberg, P.E.M. Siegbahn and M. Svensson, **32**, 4218 (1993) Inorg. Chem
34. M.R.A. Blomberg, P.E.M. Siegbahn and M. Svensson, J. Phys. Chem. **97**, 2564 (1993)
35. P.E.M. Siegbahn, Theor. Chim. Acta **86**, 219 (1993)
36. P.E.M. Siegbahn, Chem. Phys. Letters **201**, 15 (1993)
37. P.E.M. Siegbahn, Theor. Chim. Acta **87**, 441 (1994)
38. P.E.M. Siegbahn, J. Am. Chem. Soc. **115**, 5803 (1993)
39. P.E.M. Siegbahn, Theor. Chim. Acta. **87**, 277 (1994)
40. P.E.M. Siegbahn, Chem. Phys. Letters **205**, 290 (1993)
41. M.R.A. Blomberg, C.A.M. Karlsson and P.E.M. Siegbahn, J. Phys. Chem. **97**, 9341 (1993)
42. P.E.M. Siegbahn, J. Phys. Chem. **97**, 9096 (1993)
43. (a) S.R. Langhoff, L.G.M. Pettersson and C.W. Bauschlicher, Jr., J. Chem. Phys. **86**, 268 (1987); (b) C.W. Bauschlicher, Jr., S.R. Langhoff, H. Partridge and L.A. Barnes, J. Chem. Phys. **91**, 2399 (1989); (c) M. Rosi, C.W. Bauschlicher, Jr., S.R. Langhoff and H. Partridge, J. Phys. Chem. **94**, 8656 (1990); (d) M. Rosi, C.W. Bauschlicher, Jr. Chem. Phys. Lett. **166**, 189 (1990); (e) C.W. Bauschlicher, Jr. and S.R.Langhoff, J. Phys. Chem. **95**, 2278 (1991); (f) M. Sodupe, C.W. Bauschlicher, Jr., S.R. Langhoff and H. Partridge, J. Phys. Chem. **96**, 2118 (1992)
44. (a) K. Balasubramanian and C. Ravimohan, Chem. Phys. Lett. **145**, 39 (1988); (b) K. Balasubramanian and C. Ravimohan, J. Phys. Chem. **93**, 4490 (1989); (c) K. Balasubramanian and J.Z. Wang, J. Chem. Phys. **91**, 7761 (1989); (d) K. Balasubramanian and D.-W. Liao, J. Phys. Chem. **92**, 6259 (1988); (e) J.Z. Wang, K.K. Das and K. Balasubramanian, Mol. Phys. **69**, 147 (1990); (f) J. Li and K. Balasubramanian, J. Phys. Chem. **94**, 545 (1990); (g) K. Balasubramanian, P.Y. Feng and M.Z. Liao, J. Chem. Phys. **88**, 6955 (1988)
45. E.A. Carter and W.A. Goddard III, J. Phys. Chem. **92**, 5679 (1988)
46. M. Blomberg, U. Brandemark, L. Pettersson and P. Siegbahn, Int. J. Quantum. Chem. **XXIII**, 855 (1983)
47. M.R.A. Blomberg, U. Brandemark and P.E.M. Siegbahn, J. Am. Chem. Soc. **105**, 5557 (1983)
48. (a) J.J. Low and W.A. Goddard III, J. Am. Chem. Soc. **106**, 8321 (1984); (b) J.J. Low and W.A. Goddard III, Organometallics **5**, 609 (1986); (c) J.J. Low and W.A. Goddard III, J. Am. Chem. Soc. **106**, 6928 (1984); (d) J.J. Low and W.A. Goddard III, J. Am. Chem. Soc. **108**, 6115 (1986)
49. N. Koga and K. Morokuma, Chem. Rev. **91**, 823 (1991)
50. P.J. Hay, In Transition Metal Hydrides, A. Dedieu, Ed., VCH Publishers (1992), New York, pp. 127–147
51. (a) K. Kitaura, S. Obara and K. Morokuma, J. Am. Chem. Soc. **103**, 2891 (1981); (b) S. Obara, K. Kitaura and K. Morokuma, J. Am. Chem. Soc. **106**, 7482 (1984)

52. J.O. Noell and P.J. Hay, J. Am. Chem. Soc. **104**, 4578 (1982)
53. N. Koga and K. Morokuma, J. Phys. Chem. **94**, 5454 (1990)
54. T. Ziegler, V. Tschinke, L. Fan and A.D. Becke, J. Am. Chem. Soc. **111**, 9177 (1989)
55. R. Ahlrichs, P. Scharf and C. Erhardt, J. Chem. Phys. **82**, 890 (1985)
56. D.P. Chong and S.R. Langhoff, J. Chem. Phys. **84**, 5606 (1986)
57. GAUSSIAN 92, Revision A, M.J. Frisch, G.W. Trucks, M. Head-Gordon, P.M.W Gill, M.W. Wong, J.B. Foresman, B.G. Johnson, H.B. Schlegel, M.A. Robb, E.S. Replogle, R. Gomperts, J.L. Andres, K. Ragavachari, J.S. Binkley, C. Gonzales, R.L. Martin, D.J. Fox. D.J. Defrees, J. Baker, J.J.P. Stewart and J.A. Pople, Gaussian, Inc. (1992), Pittsburgh PA
58. P.J. Hay and W.R. Wadt, J. Chem. Phys. **82**, 299 (1985)
59. P.E.M. Siegbahn and M. Svensson, to be published
60. S.A. Mitchell, M.A. Blitz, P.E.M. Siegbahn and M. Svensson, J. Chem. Phys. **100**, 423 (1994)
61. M.R.A. Blomberg, U.B. Brandemark, P.E.M. Siegbahn, J. Wennerberg and C.W. Bauschlicher, Jr., J. Am. Chem. Soc. **110**, 6650 (1988)
62. T. Lee, private communication
63. C.M. Rohlfing and R.L. Martin, Chem. Phys. Lett. **115**, 104 (1985)
64. L.G.M. Pettersson and H. Åkeby, J. Chem. Phys. **94**, 2968 (1991)
65. R.L. Martin, J. Phys. Chem. **87**, 750 (1983). See also R.D. Cowan and D.C. Griffin, J. Opt. Soc. Am. **66**, 1010 (1976)
66. M.R.A. Blomberg, P.E.M. Siegbahn and M. Svensson, to be published
67. C.W. Bauschlicher, Jr. and S.R.Langhoff, Chem. Phys. Lett. **177**, 133 (1991)
68. S.A. Mitchell, Ch.12 in Gas-Phase Metal Reactions, A. Fontijn, Ed., Elsevier (1992), Amsterdam
69. M.R.A. Blomberg, P.E.M. Siegbahn, U. Nagashima and J. Wennerberg, J. Am. Chem. Soc. **113**, 424 (1991)
70. (a) J.W. Kauffman, R.H. Hauge and J.L. Margrave, J. Phys. Chem. **89**, 3541 (1985); (b) J.W. Kauffman, R.H. Hauge and J.L. Margrave, J. Phys. Chem. **89**, 3547 (1985)
71. L.S. Sunderlin and P.B. Armentrout, J. Am. Chem. Soc. **111**, 3845 (1989)
72. A.K. Rappé, C.J. Casewit, K.S. Colwell, W.A. Goddard III and W.M. Skiff, J. Am. Chem. Soc. **114**, 10024 (1992)
73. P.E.M. Siegbahn and M.R.A. Blomberg, to be published
74. Crabtree, R.H. In Selective Hydrocarbon Activation: Principles and Progress, J.A. Davies, P.L. Watson, A., Greenberg and J.F. Liebman, Ed., VCH Publishers (1990), New York, pp. 1–18
75. M.R.A. Blomberg, P.E.M. Siegbahn and M. Svensson, to be published

NOBUAKI KOGA AND KEIJI MOROKUMA

ALKENE MIGRATORY INSERTIONS AND C-C BOND FORMATIONS

1. Introduction

Alkene migratory insertion (equation *1*) is a key elementary step in various catalytic cycles [1].

(1) R–M ······ ‖ ⟶ M–(CH₂CH₂)–R

Insertion into an M-H bond, an essential step in many catalytic cycles such as hydrogenation, converts an alkene into an alkyl group. Insertion of an alkene into an M-C bond results in C-C bond formation and is a key reaction in the chain-growth step of varieties of olefin polymerization processes such as Ziegler-Natta polymerization. Because of their importance, model alkene insertion reactions have been theoretically extensively studied as will be discussed in this chapter, in order to clarify their reaction mechanisms and to obtain understanding of electronic factors that control the reactions. In this chapter we first discuss alkene coordination to a transition metal complex. In sections 3 and 4 the alkene migratory insertion into an M-H and an M-C bond, respectively, will be discussed. In section 5 theoretical studies of related insertions are shown. The last section is the summary.

2. Alkene coordination

First, we consider ethylene coordination to a transition metal complex. It is generally considered that alkene insertion takes place through a prior coordination of an alkene to the metal center [1]. Coordination of an alkene bearing

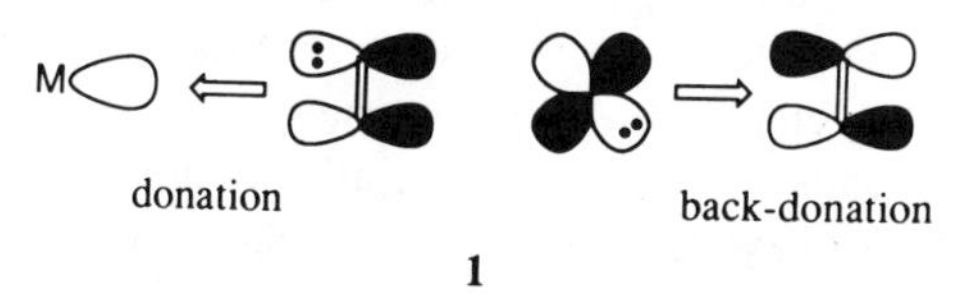

1

P.W.N.M. van Leeuwen et al. (eds), Theoretical aspects of homogeneous catalysis, 65–91.

π and π^* orbitals to a transition metal having d orbitals has been traditionally explained by so-called donation and back-donation shown in **1**.
Kitaura, Sakaki, and Morokuma analyzed the interaction energy between ethylene and $Ni(PH_3)_2$ at the *ab initio* restricted Hartree-Fock (RHF) level [2]. In this ethylene complex, the late transition metal is electron-rich with a formal electron count of d^{10}. In addition, the d orbital from which back-donation takes place is destabilized by two PH_3 groups. Consequently, back-donation has been identified to be responsible for ethylene coordination. In this study, the CC bond length in $(C_2H_4)Ni(PH_3)_2$ was calculated to be 1.42 Å. Morokuma and Borden obtained at the RHF level the CC distance of 1.44 Å in the isoelectronic Pt complex, $(C_2H_4)Pt(PH_3)_2$ [3]. These CC bond distances are about 0.1 Å longer than that in free ethylene calculated at the same level. The CC bond is stretched as a result of back-donation to anti-bonding π^* orbital. On the other hand, the CC distances in different ethylene complexes of the same group 10 transition metals, $(H)_2M(PH_3)(C_2H_4)$ (M = Ni, Pd, Pt), were calculated to be 1.34–1.35 Å, not much stretched [4, 5]. In these complexes with higher oxidation numbers and thus fewer d electrons, donation rather than back-donation is responsible for ethylene coordination. With the contribution of only the donation the CC bond cannot be stretched much. If the π^* orbital is stabilized by bending an alkene, back-donation takes place more strongly, to stretch the CC bond more [2, 3]. For instance, in $(C_2H_4)Pt(PH_3)_2$ when the bending angle of ethylene is assumed to be the same as in the strained alkene complex **2**, the CC distance is calculated to be 1.52 Å, 0.08 Å longer than that in the true $(C_2H_4)Pt(PH_3)_2$ [3].

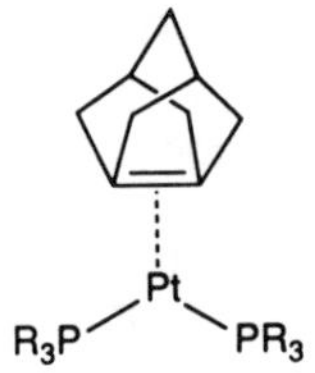

Pt
R_3P PR_3

2

In $(C_{60})Pt(PH_3)_2$ almost one full electron is transferred from the Pt fragment to the electronegative C_{60}, resulting in a long CC bond of 1.495 Å [6]. These complexes, depending on the extent that the CC π bond is broken, have some metallacyclopropane-like character.

Stretched CC bonds are also found in alkene and alkyne complexes of early transition metals having low oxidation number. Sodupe and Bauschlicher have investigated $M(C_2H_2)^+$ (M = Sc, Ti, V, Y, Zr, Nb, and Mo), to find that the calculated CC distances in these complexes are longer, compared with those in late transition metal complexes [7]. Sterigerwald and Goddard have found that the CC bond distance in $Cl_2Ti(C_2H_4)$ is 1.46 Å [8]. Koga and Morokuma have found that in the isoelectronic $Cl_2Zr(C_2H_4)$ and $Cp_2Zr(C_2H_4)$ the CC distances are 1.48 and 1.50 Å, respectively, at the RHF level [9]. These CC distances are close to that of cyclopropane, and these complexes were shown to be metallacyclopropanes rather than alkene complexes [8, 9]. Because of the small electronegativity of early transition metals in these "ethylene complexes", strong back-donation takes place to break the CC π bond and form the M-C σ bond. Accordingly, alkene coordination to X_2M (X = Cl and Cp and M = Ti and Zr) with the formal electron count of d^2 is not a simple coordination but should be taken as an oxidative addition to give a d^0 complex.

3. Alkene migratory insertion into an M-H bond

3.1. ELEMENTARY REACTIONS

By means of the *ab initio* RHF method, Koga, Morokuma, and their coworkers have studied ethylene insertion from d^8 square planar ethylene hydride complexes (equation *2*) [10].

(2) $H{-}M(C_2H_4)(H){-}PH_3 \longrightarrow [H_2C{=}CH_2 \cdots H \cdots M(H){-}PH_3]^{TS} \longrightarrow H_2C{-}CH_2(H \cdots M)(H){-}PH_3$ (M=Ni,Pd,Pt)

TS

The optimized structures of the reactant, the product, and the transition state (TS) for reaction *2* with M = Pd and Pt are shown in Figure *1*.

Alkene insertion passes through a four-centered TS with a small activation energy. At the four-centered TS, the Pd-C bond distance is short and the Pd-H bond to be broken is stretched only by 0.05 Å, indicating a tight TS. As will be shown later in this chapter, migratory insertions generally proceed through such a tight four-centered TS. At the four-centered TS, bond exchange is facilitated by the orbital interaction shown in **3**, in which an occupied σ_{MH} and a vacant $\sigma_{MH}{}^*$ orbital interact with π^* and π orbital of olefin, respectively.

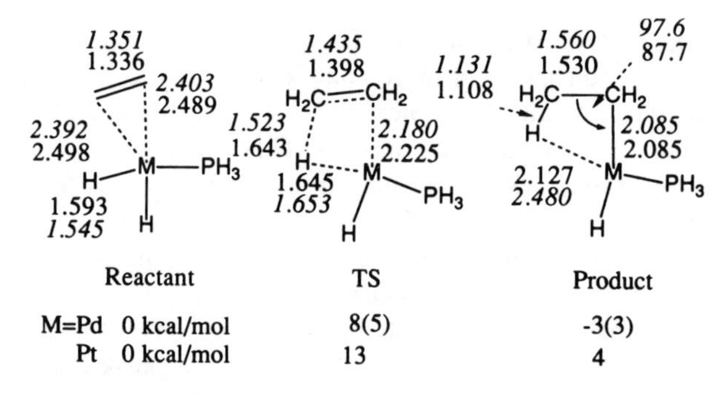

Fig. 1. RHF optimized structures (in Å and deg) for ethylene insertion into Pd-H bonds. Italics are for M = Pt. Energies are relative to the reactant at the RHF level. Numbers in parentheses are at the MP2 level.

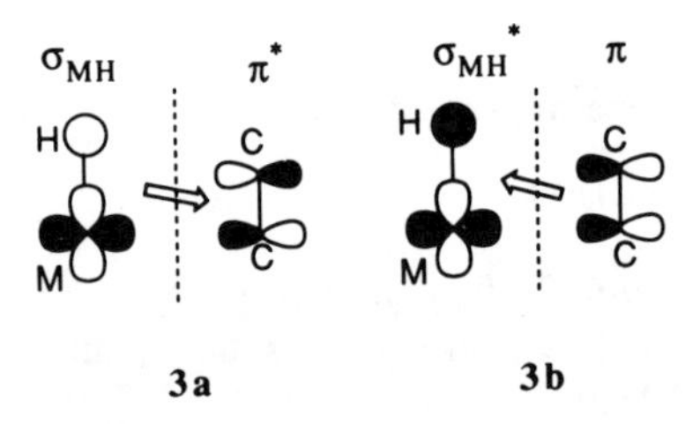

When $d_{x^2-y^2}$ shown in **4** is vacant, the σ_{MH} and σ_{MH}* orbitals have large weights of the metal d orbital, favoring interactions **3a** and **3b** because of the appropriate nodal properties. If $d_{x^2-y^2}$ is occupied, it does not play a dominant role in the σ M-H bond; σ_{MH} and σ_{MH}* orbitals would have the character of a metal s orbital rather than a metal d orbital. This makes interactions **3a** and **3b** weaker. Therefore, $d_{x^2-y^2}$ should be vacant.

4

Generally in the d^8 square planar complex such as the present $(H)_2Pd(PH_3)(C_2H_4)$, $d_{x^2-y^2}$ is formally vacant. The same conclusion was obtained in previous theoretical calculations as well as from frontier orbital arguments [11]. If $d_{x^2-y^2}$ is vacant, ethylene coordinates to the metal mainly through donation, which smoothly changes to the orbital interaction **3b** at the TS. Though back-donation is weak in this d^8 ethylene complex, the interaction **3a** corresponding to back-donation may take place at the TS where ethylene with the long CC distance has stable π^* orbital and thus back-donation is easy.

In the product Pd ethyl complex, structural features of agostic interaction, [12] an intramolecular CH . . . M interaction, are found: a long CH bond of 1.13Å, a short Pd . . . H distance of 2.13Å and a small PdCC angle of 88°. While a three-center two-electron bond has been proposed experimentally as the basis of the M . . . CH interaction, [13] Koga, Kitaura, Obara and Morokuma have found that this interaction is caused by electron donation from the βCH bond to an empty d orbital of the central transition metal, **5** [10a].

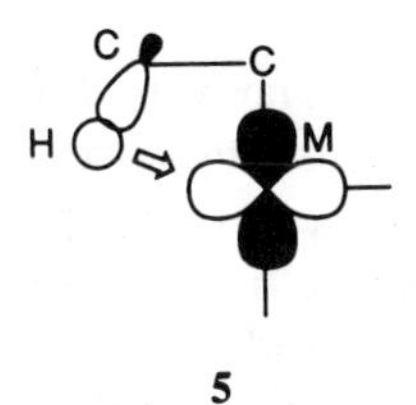

5

They have also discussed the effect of this interaction on β-hydrogen elimination, the reverse reaction of olefin insertion, finding that it incipiently activates the CH bond to lower the activation energy for β-hydrogen elimination. The interaction taking place at the TSs of alkene insertions will be discussed later.

Alkene insertion in a d^8 five-coordinate complex is assumed to take place in several catalytic cycles [14] and thus has been subject of several theoretical studies. This is more complicated than that of the d^8 four-coordinate alkene complex, although the reactant ethylene hydride complex has the same formal electron count; the Berry pseudorotation (BPR) [15] may isomerize the reactant with a low activation barrier. Thorn and Hoffmann (TH) [16] studied, with the extended Hückel (EH) method, the alkene insertion of $HPtCl(PH_3)_2(C_2H_4)$. They have shown that the most stable isomer should have a trigonal bipyramidal (TBP) structure with axial H and equatorial, in-plane ethylene, **6**.

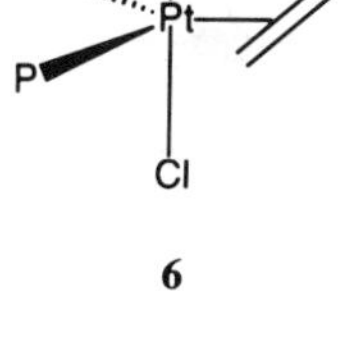

6

Since the Pt-H bond is not coplanar with the CC π bond, this structure is not suitable for alkene insertion. Their analysis shows that **7a** and **7b** are the structures suitable for the insertion.

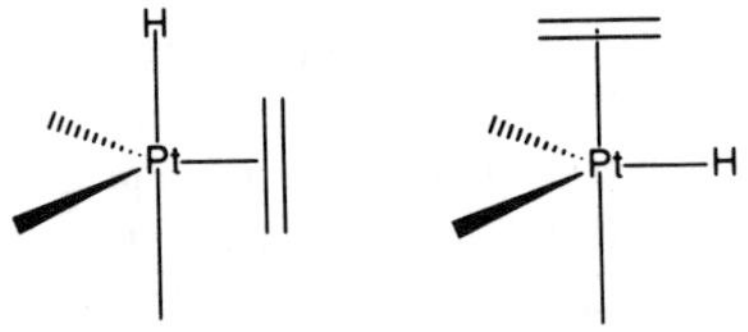

7a **7b**

The vacant d orbital of **7a** and **7b** is not very different from the d orbital in **4**. However, the activation energy from these structures calculated by the EH method is too high, over 40 kcal/mol. TH have looked for an easier reaction path and finally argued that alkene insertion from a square pyramidal (SP) isomer with the cis basal ethylene and hydride is virtually identical with that from a d^8 four-coordinate complex discussed previously.

Koga, Jin, and Morokuma (KJM) have studied the BPR process as well as the ethylene migratory insertion (equation *3*) with the *ab initio* RHF method [17] and have demonstrated that ethylene insertion of the d^8 five-coordinate complex takes place from the most stable reactant with the apical hydride and the equatorial, in-plane ethylene via a square pyramidal TS.

(3) $HRh(CO)_2(PH_3)(C_2H_4)$, **8** $\rightarrow$ $C_2H_5Rh(CO)_2(PH_3)$, **9**

They have found that, when two TBP isomers are energetically close to each other, the activation energy for BPR connecting them is quite small. The square pyramidal TSs they have found for ethylene insertion are shown in Scheme *1* and Figure *2* [17].

The skeleton of this square pyramidal TS in Figure *2* is similar to that of the TS for BPR in the same figure. This suggests that ethylene insertion is connected with BPR. At this square pyramidal TS the d orbital, **10**, is vacant and thus can play the same role as $d_{x^2-y^2}$ in the d^8 square planar complex.

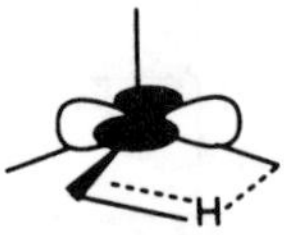

10

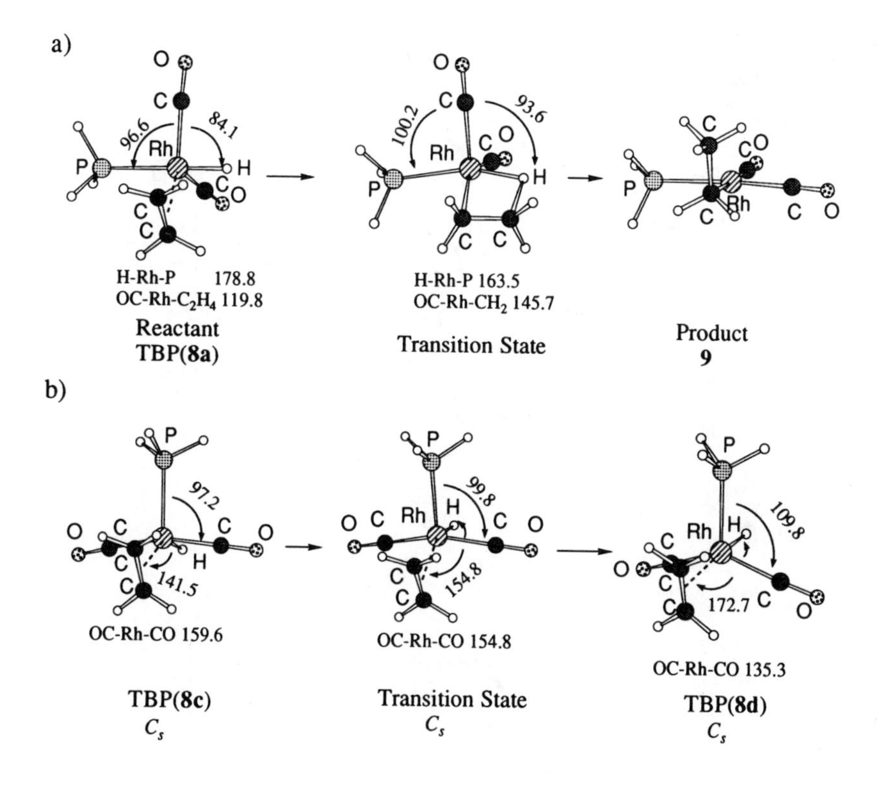

Fig. 2. RHF optimized structures (in Å and deg) of reactant, product, and transition state for (a) ethylene insertion and (b) Berry pseudorotation of $HRh(CO)_2(PH_3)(C_2H_4)$, **8.**

The reactants were shown to be the stable isomers with apical hydride and equatorial ethylene. While apical hydride can use the vacant d_z2 orbital shown in **11** to form a strong Rh-H bond, equatorial ethylene is favorable because of strong π back-donation [17, 18].

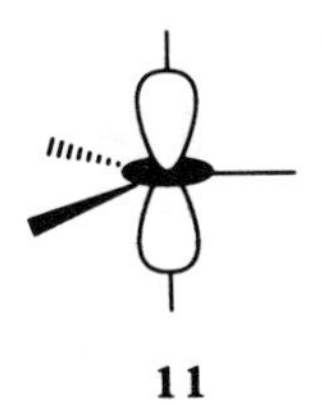

11

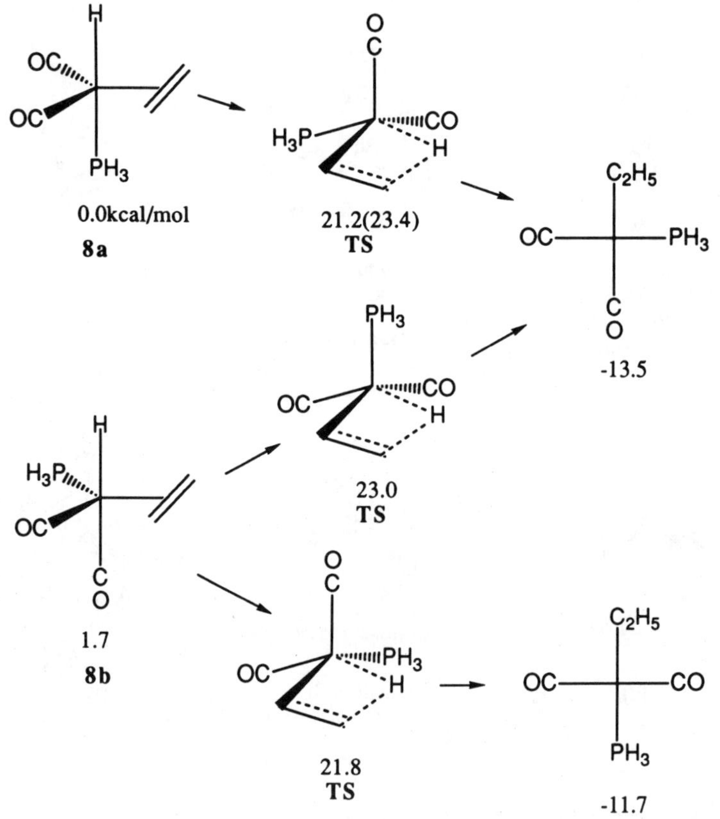

H, OC, OC, PH3, 0.0kcal/mol, **8a**; O C, H3P, CO, H, 21.2(23.4) **TS**; C2H5, OC, PH3, C O, -13.5; PH3, OC, CO, H, 23.0 **TS**; H, H3P, OC, C O, 1.7 **8b**; O C, OC, PH3, H, 21.8 **TS**; C2H5, OC, CO, PH3, -11.7

Scheme 1. Potential energy profile for ethylene insertion of $Rh(H)(CO)_2(PH_3)(C_2H_4)$, **5**. Numbers are RHF (MP2) energies relative to **5a**.

A different ethylene insertion from d^8 five-coordinate complex, reaction *4*, was studied with the *ab initio* RHF and single and double excitation configuration interaction (SDCI) method by Antolovic and Davidson (AD) [19] and with the density functional theory (DFT) by Versluis [20].

$$(4) \qquad HCo(CO)_3(C_2H_4) \rightarrow C_2H_5Co(CO)_3$$

While they have optimized the structures of the reactant and the product, they follow the reaction path assuming that all the geometrical parameters vary linearly between the reactant and the product, i.e., using the linear synchronous transient (LST) approximation. Both studies have shown that **12** is the favorable reactant structure for alkene insertion and the activation energy from it is not very large, although between these two studies there are several quantitative discrepancies including the stability of the isomers of the reactant, the activation barrier and the geometrical parameters.

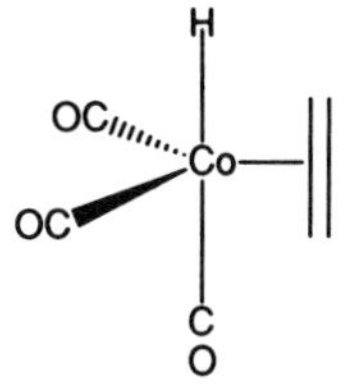

12

The LST path does not go through a square pyramidal structure, which is the TS in reaction *3*. Therefore, one will be required to relax the symmetry constraint imposed by the LST approximation to take the BPR into account and obtain a correct transition state.

In the examples discussed above, alkene has coordinated to intermediate transition metal complexes. A hydrozirconation, alkene and alkyne insertion into the Zr-H bond of $Cp_2Zr(Cl)H$, [21] shows a different potential energy profile. Endo, Koga, and Morokuma (EKM) have studied the potential energy profile of model hydrozirconation reactions (equation *5*) [22].

(5a) $Cp_2Zr(Cl)H + C_2H_4 \rightarrow Cp_2Zr(Cl)(C_2H_5)$

(5b) $Cp_2Zr(Cl)H + C_2H_2 \rightarrow Cp_2Zr(Cl)(C_2H_3)$

They have compared, with the RHF geometry optimizations and second order Møller-Plesset perturbation (MP2) energy calculations at the RHF structures, two reaction paths shown in Scheme *2*. In path 1 ethylene attacks Zr between the chloride and the hydride ligands and in path 2 from the opposite side of the chloride ligand.

They have found that path 1 requires a lower activation barrier and is more exothermic. The lower barrier was ascribed to the difference in deformation energy of the Zr complex between two paths. The distortion of the Zr complex in path 1 is easier than in path 2, since the repulsion between the hydride and the chloride ligand is smaller and the Zr-H bond partially formed at the TS for path 1 is stronger than that for path 2. In these reactions no ethylene complex has been found, different from the cases of late transition metal

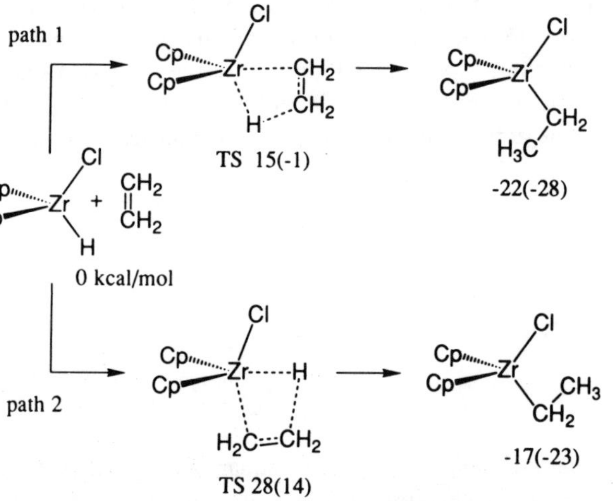

path 1

TS 15(-1)

-22(-28)

0 kcal/mol

path 2

TS 28(14)

-17(-23)

Scheme 2. Potential energy profile for two possible reaction paths of hydrozirconation. Energies are relative to reactants in kcal/mol at the RHF (MP2//RHF) level.

complexes shown above. There are two possible reasons. One is that, though ethylene coordination would be possible through donation even in the neutral Zr complex with the formal electron count of d^0, the small electronegativity of Zr prevents donation. Ethylene coordination would require a positive charge on Zr, which increases the electronegativity of Zr. The other reason must be the steric repulsion between ethylene and the Cp ligands.

They have also studied hydrozirconation of acetylene, to show that it is intrinsically slightly more difficult than that of ethylene. This is not in agreement with the experimental results that alkynes are hydrozirconated much faster than some alkenes [23]. These alkenes are substituted alkenes, and they attribute this discrepancy to the steric repulsion between the Cp ligands and the substituents on alkenes, which would increase the activation barrier.

The steric repulsion between the Cp ligands and alkene was found to be important also in comparing the present insertion of ethylene into the Zr-H bond of $Cp_2Zr(Cl)(H)$ with that of Cl_3ZrH [22]. In the latter reaction, while path 1 is down-hill and more favorable than path 2, there exists an ethylene complex in path 2 and the activation energy from this complex was calculated to be 4 kcal/mol. Path 2 with the Cl ligands is easier than that with the Cp ligands; the replacement of Cp by Cl makes it easier for ethylene to approach the Zr complex through path 2. Consequently, EKM has reasoned that the larger size of the Cp ligands make path 2 definitely less favorable than path 1.

Furthermore, EKM have compared the potential energy profiles for insertion of ethylene into the Zr-C bond of $Cp_2Zr(Cl)(CH_3)$ with those for hydrozirconation above.

$$(6) \qquad Cp_2Zr(Cl)(CH_3) + C_2H_4 \rightarrow Cp_2Zr(Cl)(C_3H_7)$$

They have found that insertion into the Zr-C bond is quite unfavorable with a high activation energy and have ascribed this to the directionality of sp^3 hybrid of the methyl group. This high activation barrier will later be compared with the low activation barrier for propagation step of polymerization.

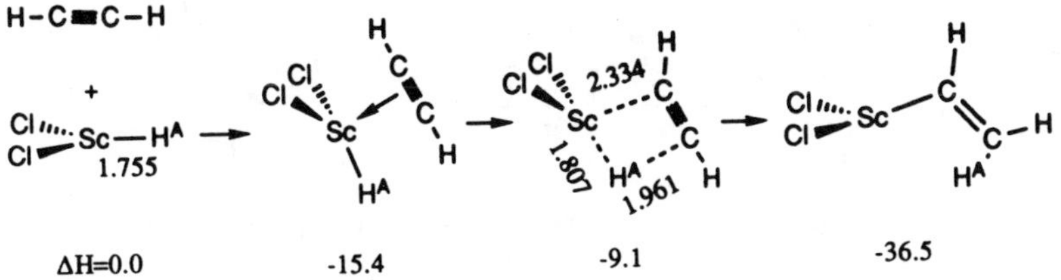

Fig. 3. Energy profiles in kcal/mol at the CI level for acetylene insertion to Cl_2ScH.

Another example of insertion of an early transition metal complex theoretically studied is reaction *7*.

$$(7) \qquad Cl_2ScH + C_2H_2 \rightarrow Cl_2ScC_2H_3$$

Experimentally, in the reaction of propyne with Cp^*_2ScR (R = H, CH_3) only σ-metathesis, in competition with insertion, takes place [24]. Rappe has studied the reaction of Cl_2ScH with acetylene with the generalized valence bond (GVB) and the CI method in order to compare these two reaction paths [25]. As shown in Figure *3* he has found that the activation energy of 6 kcal/mol for the insertion from the π-complex is comparable to that of σ-metathesis of acetylide formation. Thus the computational results are not consistent with the experiment.

Very recently, Siegbahn have studied ethylene insertion into an M-H bond of MH_n (n = 1–3) for the second row transition metals with the RHF geometry optimization and modified coupled pair functional (MCPF) energy calculations. Though these reactions are not directly related to elementary steps of any catalytic cycle, their study provides some basic understanding of alkene insertion. He has analyzed the difference in energetics between the metals in terms of promotion energy, exchange energy, sd-hybridization, and metal-ligand repulsion. Qualitative conclusion is that repulsion between non-bonding metal electrons and alkene electrons are important part of the activation barrier, which can be reduced by sd-hybridization and promotion to an s^0 state. Additional hydrogen ligands were shown to attract metal electrons to reduce the repulsion and thus the activation barrier.

3.2. CATALYTIC CYCLE INCLUDING ALKENE MIGRATORY INSERTION – HYDROGENATION

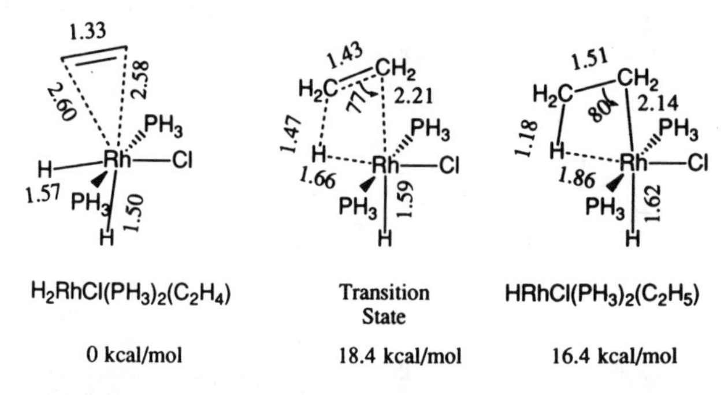

Fig. 4. RHF optimized structures (in Å and deg) for ethylene insertion of $H_2RhCl(PH_3)_2(C_2H_4)$. Energies are relative to the reactant in kcal/mol.

Alkene insertion, equation *8*, is a model reaction of alkene insertion step in hydrogenation catalytic cycle by Wilkinson catalyst and has been experimentally considered to be rate-determining [26].

$$(8) \qquad H_2RhCl(PH_3)_2(C_2H_4) \rightarrow HRhCl(PH_3)_2(C_2H_5)$$

Dedieu and his coworkers [27] have made a very early study on the potential energy profile of this catalytic cycle using experimental or assumed structures at the RHF level. More recently, Daniel, Koga, Han, Fu and Morokuma (DKHFM) have obtained the potential energy profile of this catalytic cycle by determining structures of the transition states as well as the intermediates for all the elementary steps by the *ab initio* RHF energy gradient method [28]. The results for this model alkene insertion by DKHFM is shown in Figure *4*.

The reactant of reaction *8* has a vacant d orbital, similar to **3** and appropriate for the interaction at the four-centered TS mentioned above. However, the activation barrier for reaction *8* was calculated to be much higher, because of a large endothermicity. This large endothermicity was ascribed to the strong Rh-H bond in the reactant, which was to be broken. Since the ligand *trans* to H is Cl, which has a weak *trans* influence, the Rh-H bond is strong. Furthermore the strongly electron-donating H and ethyl are *trans* to each other in the product. These factors result in large endothermicity and high activation energy.

It is necessary to investigate the potential energy profile of the entire catalytic cycle, in order to see whether this step is actually rate-determining. The model catalytic cycle studied by DKHFM is shown in Scheme *3*.

This cycle consists of (i) H_2 oxidative addition to the model catalytic active intermediate, $RhCl(PH_3)_2$, (ii) olefin coordination, (iii) olefin insertion just discussed, (iv) isomerization of alkyl hydride complex, and (v) reductive elimination which produces alkane and regenerates the catalytic active intermediate, $RhCl(PH_3)_2$. This catalytic cycle was originally proposed by Halpern *et al.*, [26] though isomerization of the alkyl hydride complex was not included in the original mechanism. The resultant potential energy profile is shown in Figure *5*.

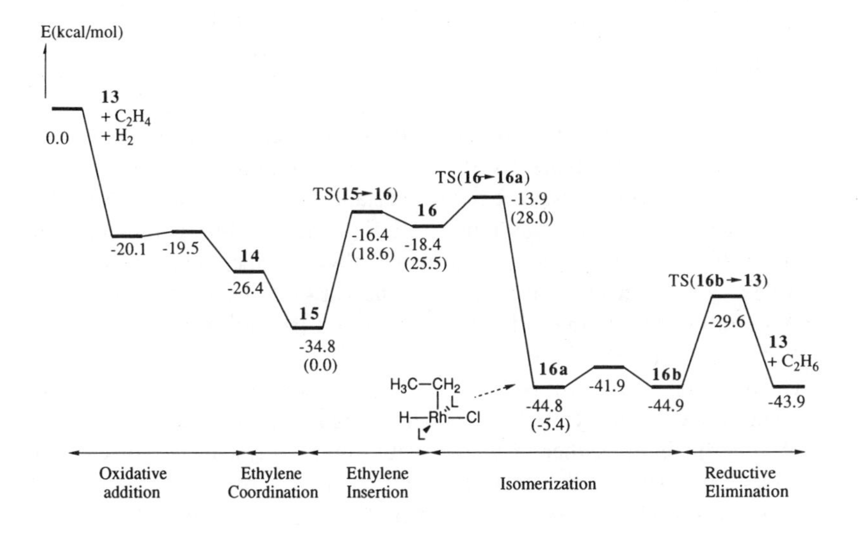

Fig. 5. Potential energy profile of the entire catalytic cycle of the Halpern mechanism of olefin hydrogenation by the Wilkinson catalyst, in kcal/mol at the RHF level, relative to **13**+C_2H_4+H_2. Numbers in parenthesis are the MP2 energy at the RHF structures, relative to **15**.

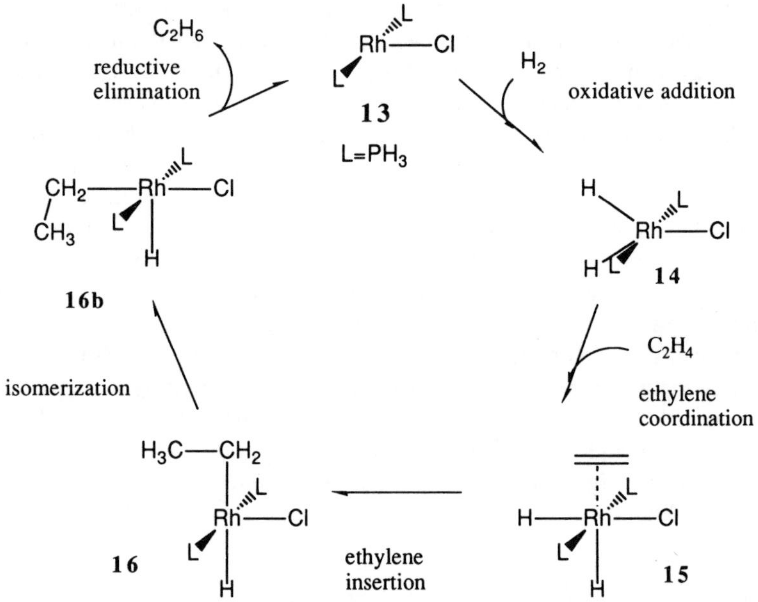

Scheme 3.

discussed takes place. The subsequent isomerization with a low activation barrier gives a more stable cis ethyl hydride complex. The final step of reductive elimination takes place with a substantial activation energy. Actually, the rate-determining step was found to be alkene insertion combined with the first step of isomerization of the *trans* ethyl hydride complex to the cis complex with an effective overall barrier height of 21 kcal/mol. Their theoretical study suggests that alkene insertion and isomerization of the ethyl hydride complex may take place as one step.

The potential energy profile which DKHFM constructed from the energetics of these five elementary steps is smooth without excessive barriers and too stable intermediates, either of which would break the sequence of elementary reactions. Though this feature would be what is expected in any good catalyst, this study has provided the first theoretical demonstration that it actually is so in a model cycle.

When the alkene insertion is almost thermoneutral as shown in the case of the Pd and Pt insertion reactions, it can proceed with a low activation energy.

TABLE I

Energy of reaction (ΔE) and activation energy ($\Delta E^{\dagger}$) relative to ethylene complex (in kcal/mol) for ethylene insertion into the M-CH_3 bond of $X_2MCH_3^+$.[a]

X_2	symmetry	method	$\Delta E^{\dagger}$	ΔE	Ref.
M = Ti					
Cl_2	C_s	RHF	11		31
Cl_2[b]	C_s	RHF	12	−2	32
Cl_2	C_1	RHF	14	−3	33
		RMP2	6	–[c]	33
		DPUMP2[d]	4	−11	33
Cp_2[b]	C_s	RHF	22		32
		RMP2	10	−12	32
M = Zr					
Cl_2	C_s	RHF	22	9	37
		CVB-CI	24	13	37
H_2SiCp_2	C_1	RHF	17	−5	34
		RMP2	6	−5	34

[a]Optimized at the RHF level.
[b]Optimized with the PRDDO approximation.
[c]Second order perturbation method could not be applied.

DKHFM have found that, when the chloride of the Wilkinson catalyst is replaced by a hydride, ethylene insertion is almost thermoneutral, since the hydride weakens the Rh-H bond to be broken in the insertion. Thus, they have discussed the effect of the transition metal and the ligand using model reactive intermediate, $RhH(PH_3)_2$, $Pt(PH_3)$, and $Pd(PH_3)$. In the cycle of $Pt(PH_3)$, olefin insertion from $H_2Pt(PH_3)(C_2H_4)$ requires a smaller activation energy as shown above. Though it seems that the barrier for alkene insertion step for $RhH(PH_3)_2$ and $Pt(PH_3)$ is low and thus the catalytic cycle could proceed easily, one has to note that reverse β-hydride elimination can also take place with a low activation energy. These results have suggested that $RhH(PH_3)_2$ and $Pt(PH_3)$ are good catalysts for alkene isomerization rather than for alkene hydrogenation. On the other hand, the high activation energy for reverse β-hydride elimination of the Wilkinson catalyst prevents alkene isomerization. Chloride in the Wilkinson catalyst seems to play an important role in making the insertion rate-determining.

4. Alkene migratory insertion into an M-R bond

Alkene migratory insertion into an M-R bond, called carbometalation of alkenes, also satisfy the orbital interaction, **3**, and therefore is expected to take place through a four-centered TS. This reaction is very important, since it is a key reaction in the chain growth step of olefin polymerization such as Ziegler-Natta polymerization [1]. In addition to interest in the reaction mechanism, the control of the regio- and stereoselectivity in the alkene insertion by electronic as well as steric effects of substituents is drawing a substantial attention. High regio- and stereoselectivity is essential, for instance, in producing the highly tactic polymer and in synthesizing molecules with high e.e.

4.1. POLYMERIZATION AND TACTICITY

For the generic mechanism of polymer chain propagation step, the Cossee mechanism shown in Scheme *4* has been accepted widely as the most plausible [29].

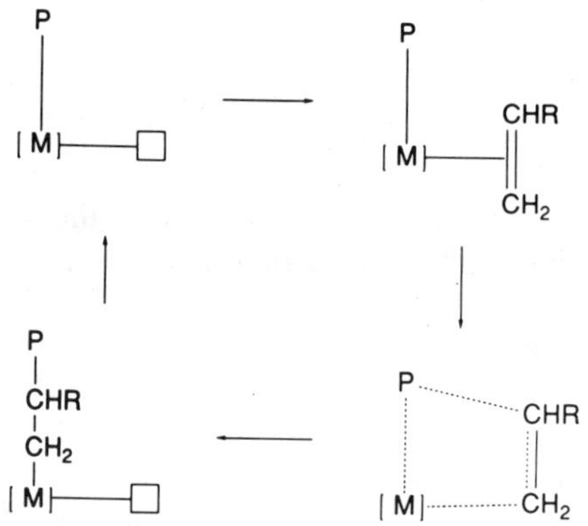

Scheme 4.

Following this mechanism, an early *ab initio* MO study of C_2H_4 + $CH_3TiCl_4AlCl_2$ by Novaro, Blaisten-Barojas, Clementi, Giunchi, and Ruiz-Vizcaya has shown that the activation energy of insertion relative to the ethylene complex is low [30]. They have not used the energy gradient nor have taken electron correlation into account. Therefore, their results should be considered to be rather qualitative. On the other hand, model reaction 9 has been studied by several groups with the energy gradient method [31–33]. The results are summarized in Table *I*.

(9) X=Cl, Cp

$X_2Ti^+CH_3 + CH_2{=}CH_2 \longrightarrow$ (π-complex) $\longrightarrow$ (four-centered TS) $\longrightarrow$ $X_2Ti^+CH_2CH_2CH_3$

Fujimoto, Yamasaki, Mizutani, and Koga have studied reaction *9* with X = Cl [31]. They have optimized the TS structure at the RHF level under the C_s symmetry constraint to show that the TS structure is four-centered as expected and that the activation energy is low (10.5 kcal/mol). They have also discussed the orbital interaction between $Cl_2TiCH_3^+$ and C_2H_4 using paired interacting orbitals to point out the importance of vacant d orbital participation just as in ethylene insertion into an M-H bond shown above.

Jolly and Marynick have studied reaction *9* with X = Cl and Cp (cyclopentadienyl) [32]. They have determined the structures at an approximate *ab initio* level with the partial-retention-of-diatomic-differential-overlap (PRDDO) approximation under the C_s symmetry constraint and have followed the reaction path using LST/partial optimization approach. The activation energy for X = Cl at the *ab initio* RHF level was calculated to be 11.9 kcal/mol. For X = Cp the activation energy calculated was 22.1 kcal/mol at the RHF level and 9.8 kcal/mol at the MP2 level.

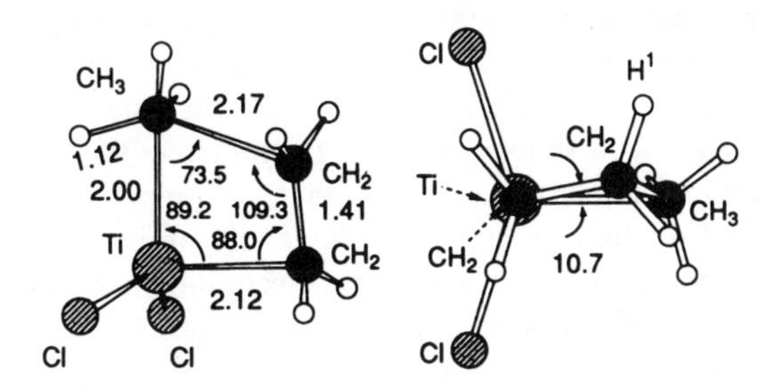

Fig. 6. RHF optimized structures (in Å and deg) of transition state for ethylene insertion into the Ti-CH_3 bond of $Cl_2TiCH_3^+$ (reaction 9).

The above two studies have shown that the polymer propagation step is quite easy in these model catalysts. However, at the C_s transition state all the CH bonds of the propyl group are eclipsed and thus this TS may be unrealistic.

Kawamura-Kuribayashi, Koga, and Morokuma (KKM) have studied reaction 9 with X = Cl without any space symmetry restriction, finding that the four-centered TS is slightly nonplanar, presumably to avoid eclipsing of CH bonds as shown in Figure *6* [33].

The unrestricted (U)MP2 calculation with triplet and quintet projection (DPUMP2) gave the activation energy of 4.3 kcal/mol; the reaction from the ethylene complex can take place very easily. They included the electron correlation effect at the DPUMP2 level because the RHF wave function was unstable with respect to the UHF wave function, due to the strong covalent character of the relevant Ti-C bonds. In fact, the RMP2 energy was too low, showing that the RMP2 result is far from convergence of the perturbation series.

KKM have also studied propylene insertion by replacing one of the hydrogen atoms in ethylene by a methyl group. As shown in Scheme *5*, they have found that the primary, i.e. head-to-tail insertion is easier than the secondary insertion, in agreement with the experiment and the Markownikov rule of organic chemistry.

In addition, they have found that nonplanarity of the transition state could give rise to stereoselectivity; TS1, with a methyl group located in the less crowded region, is more stable than TS2.

KKM have later extended their studies of polymerization using more realistic model active intermediate of a homogeneous silylene-bridged zirconocene catalyst, $(SiH_2Cp_2)ZrCH_3^+$ [34]. Since Kaminsky and his coworkers have discovered highly active homogeneous metallocene catalysts with methylaluminoxane, [35] the homogeneous Ziegler-Natta catalyst has become a center of focus of new activities in this field. $(SiH_2Cp_2)ZrCH_3^+$ is a model active intermediate of the catalyst which has been found to yield highly isotactic polypropylene by Yamazaki and his co-workers [36].

The potential energy profiles of reaction *10* calculated at the RHF as well as MP2 level are also shown in Table *1*. Similar to the reaction of the Ti complex, this reaction proceeds via an ethylene complex and a four-centered TS. Different from hydrozirconation, ethylene coordinates through donation to this highly coordinatively unsaturated, electron-deficient intermediate with the binding energy of 34 kcal/mol. The activation barrier relative to the ethylene complex was calculated to be 6 kcal/mol, to show that ethylene insertion is a quite easy process. The propyl complex with a bCH agostic interaction was found to be the most stable product.

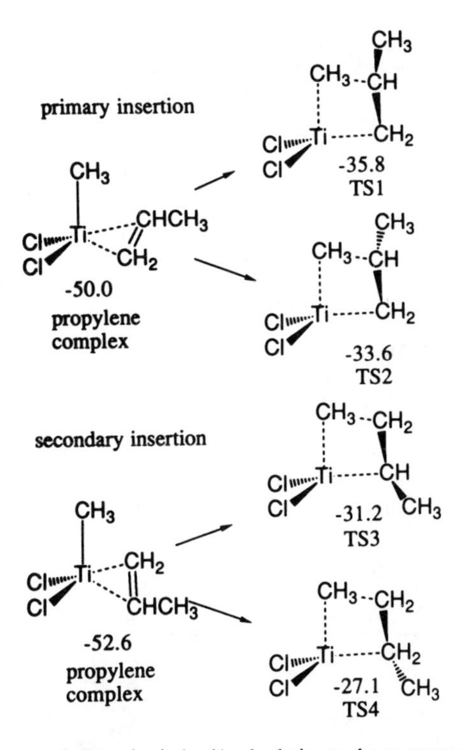

Scheme 5. Energies in kcal/mol relative to the reactants

The activation energy of 6 kcal/mol is much smaller than that for the ethylene insertion into the Z-C bond of $Cp_2Zr(Cl)(CH_3)$ discussed above. Comparing these two reactions, EKM [22] have concluded that the origin of the low activation barrier is the small deformation energy of $(SiH_2Cp_2)Zr(CH_3)^+$

(10)

TABLE II

MM Steric energies for propylene insertion into the Zr-M′ bond of $[SiH_2(CpMe_n)_2]ZrR'^+$ relative to the primary insertion with R^1 = Me

	n	primary		secondary	
		R^1 = Me	R^2 = Me	R^3 = Me	R^4 = Me
p-complex					
R′ = Me					
	0	0.0	0.0	–1.2	–1.2
	2(3,4′-Me_2)	0.0	1.1	–1.3	1.8
	4(3,5,2′,4′-Me_2)	0.0	1.2	2.4	1.7
Transition state					
R′ = Me					
	0	0.0	0.1	3.4	3.0
	2(3,4′-Me_2)	0.0	0.9	2.6	11.4
	4(3,5,2′,4′-Me_2)	0.0	0.9	4.2	11.3
R′ = Et					
	0	0.0	3.1	5.1	4.5
	2(3,4′-Me_2)	0.0	4.0	4.8	16.3
	4(3,5,2′,4′-Me_2)	0.0	3.9	9.6	19.6

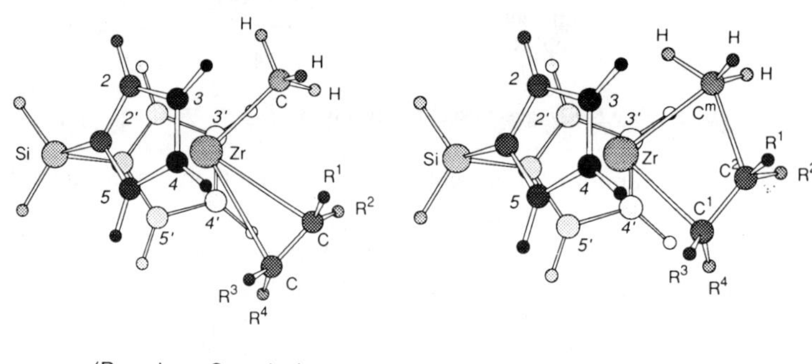

required to reach the TS. This cationic species is coordinatively unsaturated and thus easy to deform. In addition, at the TS of the model polymerization the strong agostic interaction takes place between σCH bond of the methyl group and the Zr atom, as indicated by the long CH bond of the migrating methyl group (see the CH bond length of 1.12 Å in Figure *6* for reaction *9* with X = Cl), stabilizing the TS and consequently reducing the deformation energy. On the other hand, in the neutral system where the Zr electronegativity is small, the electron donation from the CH bond to Zr, the origin of agostic interaction, does not take place [22] at the TS for ethylene insertion into the Zr-CH_3 bond of $Cp_2Zr(Cl)(CH_3)$.

KKM have further studied with molecular mechanics (MM) calculations the origin of regio- and stereoselectivity in isospecific propylene polymerization [34]. For this purpose, they have introduced alkyl substituents on the structures of the TS and the ethylene complex determined by MO calculations for reaction *10*, in order to investigate the effects of alkyl substituents on alkenes and on the Cp rings as well as the effects of the length of a model polymer chain end. Some typical examples are shown in Table *II*, where the steric energies of the four regio- and stereoisomeric TSs are listed.

The results with R′ = CH_3 show that the order of steric energy at the TSs is in good agreement with the regioselectivity, the experimental preference of primary insertion over secondary insertion, and that, if π-complexes instead of TSs are used for comparison, however, the opposite trend is found. The calculations with R′ = Et at the TSs show that one of the isomeric TSs is much more favorable. The detailed analysis has demonstrated that the substituents on the Cp rings determine the conformation of the polymer chain

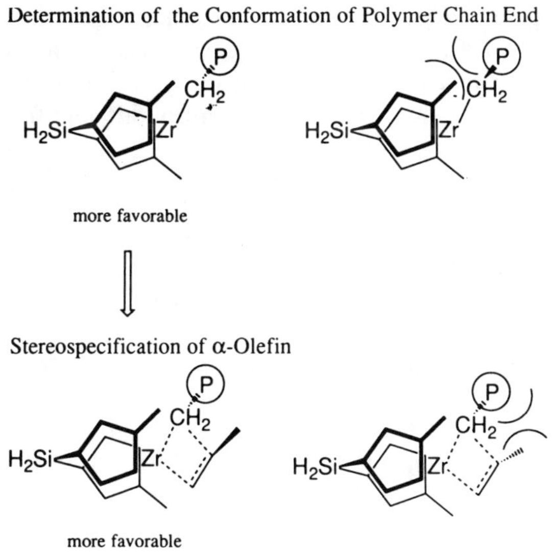

Determination of the Conformation of Polymer Chain End

P
CH_2
H_2Si
Zr

more favorable

Stereospecification of α-Olefin

P
CH_2
H_2Si
Zr

more favorable

Scheme 6.

end and that this fixed polymer chain end conformation in turn determines the stereochemistry of propylene insertion at the TS: indirect stereocontrol mechanism shown in Scheme *6*. The ethyl group is the shortest model of a polymer chain end needed to incorporate the effect of the chain end conformation. The authors have also found that the same indirect control mechanism is operational in syndiotactic polymerization.

Castonguay and Rappe have adapted a similar method to study the tacticity in propylene polymerization [37]. They have determined the structures of the stationary points for the model reaction *11* at the RHF level and carried out the energy calculations at those structures with the GVB-CI method.

$$(11) \quad Cl_2ZrCH_3^+ + C_2H_4 \rightarrow Cl_2ZrC_3H_7^+$$

Then, they have performed the MM calculations for the realistic system of (*S,S*)–C_2H_4(4,5,6,7-tetrahydro-1-indenyl)$_2$Zr(CH_3)(CH_2 = $CHCH_3$) with the fixed Zr-propylene distance corresponding to halfway along the reaction coordinate between the π-complex and the TS, to determine the structure of the "activated complex" and the favorable conformations there. They have demonstrated with the calculations of several catalysts that such calculations reproduce the tacticities experimentally observed well. They have concluded, without detailed analysis, that olefin complexation takes place with the methyl group of propylene located in the least crowded region, to determine the stereochemistry of the growing polymer chain. What is the most important factor and how it regulates the stereochemistry, which would be useful in designing catalyst, however, has not been given, different from the KKM's study.

4.2. REGIOSELECTIVITY

Nakamura, Miyachi, Koga, and Morokuma (NMKM) have studied the origin of the regioselectivity experimentally observed in carbocupration as well as carbolithiation of substituted alkynes shown in Scheme 7 [38].

Substituents having the second row atoms such as S, P, and Si give the product **17** and those bearing the first row atoms such as O and N favor the different regioisomer **18**. They have determined the structures of the stationary points for model reactions of acetylenes with OH, SH, SiH_3, and CH_3 as a substituent and analyzed the differences in activation barrier. Similar to the alkene migratory insertion, these reactions of substituted acetylenes were found to pass through a π-complex and a tight, four-centered TS. The differences in activation barriers were found to be correlated with the experimental

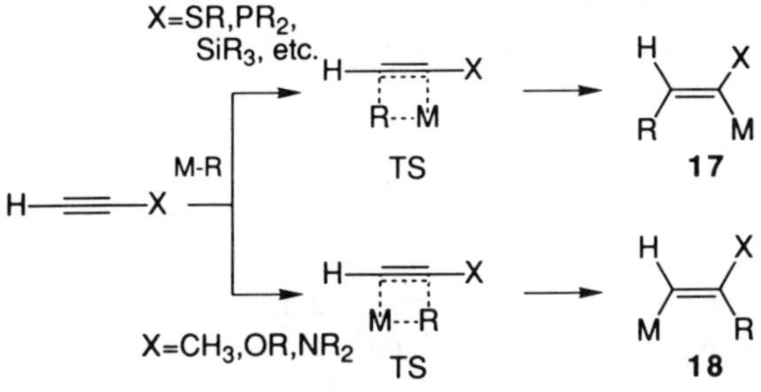

X=SR,PR$_2$, SiR$_3$, etc.

H—≡—X —(M-R)→ TS (H—≡—X / R---M) → 17 (H, X, R, M)

H—≡—X —(M-R)→ TS (H—≡—X / M---R) → 18 (H, X, M, R)

X=CH$_3$,OR,NR$_2$

Scheme 7.

regioselectivity. A detailed analysis has shown that the difference in the energy of two regioisomeric TSs is controlled by electrostatic interaction between CH_3M and substituted acetylenes, irrespective of the metal or the substituent. The results [38] of natural population analysis shown in Scheme *8* demonstrates that polarization in acetylenes with the second-row substituents is different from that with the first-row substituents, the trend in agreement with the result that the electrostatic interaction dominates the regioselectivity.

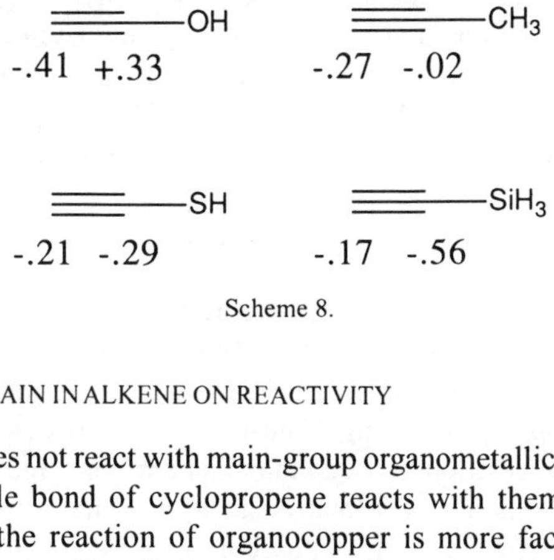

≡—OH: -.41 +.33

≡—CH$_3$: -.27 -.02

≡—SH: -.21 -.29

≡—SiH$_3$: -.17 -.56

Scheme 8.

4.3. EFFECT OF STRAIN IN ALKENE ON REACTIVITY

While ethylene does not react with main-group organometallics such as Et_2Mg, the strained double bond of cyclopropene reacts with them rapidly. It has been shown that the reaction of organocopper is more facile than that of main-group organometallics [39]. In order to investigate the effect of strain as well as the role of d orbitals of copper, Nakamura, Nakamura, Miyachi, Koga, and Morokuma (NNMKM) have studied the reactions of cyclopropene and ethylene with CH_3Cu as well with CH_3Li [40]. They have clarified that the ring strain in cyclopropene of 24 kcal/mol [41] gives a large exothermicity and is the origin of high reactivity. They have also found an interesting difference between reactions of CH_3Cu and CH_3Li. In the reaction of the latter the TS is early as expected from the large exothermicity because of the ring strain. On the other hand, the TS for CH_3Cu is asynchronous with an advanced C=C bond cleavage. This cleavage results in rehybridization of cyclopropene carbons and thus larger orbital interaction between the CC π bond and copper vacant 4p orbitals. Consequently, the authors have proposed that there are two effects of strain; the primary strain effect due to a simple strain which drives a molecule to deform to a more stable structure and the secondary strain effect where a soft metal further assists deformation by its favorable orbital interaction. On the other hand, the authors have shown that 3d orbitals of the copper do not actively participate in the bond exchange.

5. Other related insertions

There are several examples of insertions of unsaturated organic molecules into M-R bonds. Mechanistically, they are similar to the insertion of alkene, because such compounds have π and π^* orbitals as an alkene does; they pass through a four-centered TS and are intrinsically easy reactions. Theoretical studies of such insertions include that of formaldehyde into the Ru-H bond of $HRu(CO)_4$ by Nakamura and Morokuma [42] and that into the Co-H bond of $HCo(CO)_3$ by Versluis and Ziegler [43] and that of CO_2 into an Cr-H bond of $HCr(CO)_5^-$ $HCr(CO)_5$ by Bo and Dedieu [44] and into Cu-H bonds of $HCu(PH_3)_2$ and $HCu(PH_3)_3$ by Sakaki and Ohkubo [45].

6. Summary

In this chapter we reviewed recent theoretical studies of alkene migratory insertions into M-H and M-C bonds of transition metal complexes. Investigations of potential energy surfaces with *ab initio* MO methods have shown that these insertions pass through four-centered transition states. Qualitative analysis of the interaction between the alkene π bond and the M-R bond has shown that participation of formally vacant d orbitals is necessary to make the activation barrier low. This situation is easily achieved in reactions of four-coordinate d^8 ethylene hydride complexes. In the case of five-coordinate d^8 ethylene hydride complexes, the transition state is a square pyramid which provides an appropriate vacant d orbital. Early transition metal complexes also have such vacant d orbitals and thus require a low activation barrier.

An important aspect of alkene insertion is the control of regio- and stereochemistry. After the basic understanding mentioned above was achieved, attempts to investigate electronic and steric origins of these controls are being carried out in model reactions of polymerization and in some model carbometalations with the aid of molecular mechanics and MO calculations.

Nobuaki Koga
School of Informatics and Sciences and Graduate School of Human Informatics, Nagoya University, Nagoya 464–01, Japan

Keiji Morokuma
Cherry L. Emerson Center for Scientific Computation and Department of Chemistry, Emory University, Atlanta, GA 30322, USA

References

1. See for instance: (a) J.P. Collman, L.S. Hegedus, J.R. Norton, and R.G. Finke, Principles and Applications of Organotransition Metal Chemistry ; University Science Books: Mill Valley, California (1987); (b) F.A. Cotton and G. Wilkinson, Advanced Inorganic Chemistry ; Wiley: New York (1988), (c) A. Yamamoto, Organotransition Metal Chemistry ; Wiley: New York (1986)
2. K. Kitaura, S. Sakaki, and K. Morokuma, Inorg. Chem. **20**, 2292 (1981)
3. K. Morokuma and W.T. Borden, J. Am. Chem. Soc. **113**, 1912 (1991)
4. N. Koga, S. Obara, K. Kitaura, and K. Morokuma, J. Am. Chem. Soc. **107**, 7109 (1985)
5. N. Koga and K. Morokuma, in "Quantum Chemistry: The Challenge of Transition Metals and Coordination Chemistry", Nato Advanced Institute Series C; A. Veillard, Ed.; D. Reidel (1986), Dordrecht, Vol. 176, p. 351
6. N. Koga and K. Morokuma, Chem. Phys. Lett. **202**, 330 (1993)
7. M. Sodupe and C.W. Bauschlicher, J. Phys. Chem. **95**, 8640 (1991)
8. M.L. Sterigerwald and W.A. Goddard, J. Am. Chem. Soc. **107**, 5027 (1985)
9. N. Koga and K. Morokuma, unpublished results
10. (a) N. Koga, K. Kitaura, S. Obara, and K. Morokuma, J. Am. Chem. Soc. **107**, 7109 (1985); (b) N. Koga and K. Morokuma, in Quantum Chemistry: The Challenge of Transition Metals and Coordination Chemistry ; A. Veillard, Ed.; NATO ASI series, vol. 176: Reidel, Dordrecht, 1986, pp. 351
11. (a) D.L. Thorn and R. Hoffmann, J. Am. Chem. Soc. **100**, 2079 (1978); (b) S. Sakaki, H. Kato, H. Kanai, and K. Tarama, Bull. Chem. Soc. Jpn. **48**, 813 (1975); (c) K. Fukui and S. Inagaki, J. Am. Chem. Soc. **97**, 4445 (1975); (d) K. Tatsumi, K. Yamaguchi, and T. Fueno, J. Mol. Catal. 2, **437 (1977)**; (e) J.-E. Bäckvall, E.E. Björkman, L. Pettersson, and P. Siegbahn, J. Am. Chem. Soc. **106**, 4369 (1984)
12. (a) M. Brookhart and M.L.H. Green, J. Organomet. Chem. **250**, 395 (1983); (b) M. Brookhart, M.L.H. Green, and L.-T. Wong, Prog. Inorg. Chem. 36, 1 (1988)

13. F.A. Cotton, T. LaCour, and G. Stanislowski, J. Am. Chem. Soc. **96**, 734 (1974)
14. For instance, see reference 1
15. (a) I. Ugi, D. Marquarding, H. Klusacek, and P. Gillespie, Acc. Chem. Res. **4**, 288 (1971); (b) R.R. Holmes, Acc. Chem. Res. **5**, 296 (1972); (c) R.S. Berry, J. Chem. Phys. **32**, 933 (1960); (d) E.L. Muetterties, J.Am. Chem. Soc. **91**, 1636, 4115 (1969); (e) I. Ugi, F. Ramirez, D. Marquarding, H. Klusacek, G. Gokel, and P. Gillespie, P. Angew. Chem., Int. Ed. Engl. **9**, 725 (1970)
16. D.L. Thorn and R. Hoffmann, J. Am. Chem. Soc. **100**, 2079 (1978)
17. N. Koga, S.-Q. Jin, and K. Morokuma, J. Am. Chem. Soc. **110**, 3417 (1988)
18. A.R. Rossi and R. Hoffmann, Inorg. Chem. **14**, 365 (1975)
19. (a) D. Antolovic and E.R. Davidson, J. Am. Chem. Soc. **109**, 977 (1987); (b) D. Antolovic and E.R. Davidson, J. Am. Chem. Soc. **109**, 5828 (1987)
20. L. Versluis, Ph.D. Thesis, The University of Calgary, 1989
21. D.W. Hart and J. Schwartz J. Am. Chem. Soc. **96**, 8115 (1974)
22. J. Endo, N. Koga, and K. Morokuma, Organometallics, **12**, 2777 (1993)
23. D.W. Hart, T.F. Blackburn, and J. Schwartz, J. Am. Chem. Soc. **97**, 679 (1975)
24. M.E. Thompson, S.M. Baxter, A.R. Bulls, B.J. Burger, M.C. Nolan, B.D. Santarsiero, W.P. Schaefer, and J.E. Bercaw, J. Am. Chem. Soc. **109**, 203 (1987)
25. A.K. Rappe, Organometallics **9**, 466 (1990)
26. (a) J. Halpern and C.S. Wong, J. Chem. Soc., Chem. Commun. 629 (1973); (b) J. Halpern, In Organotransition Metal Chemistry ; Y. Ishii and M. Tsutsui, Eds.; Plenum: New York, 1975, pp. 109; (c) J. Halpern, T. Okamoto, and A. Zakhariev, J. Mol. Catal. **2**, 65 (1976)
27. (a) A. Dedieu, Inorg. Chem. **19**, 375 (1980); (b) A. Dedieu and I. Hyla-Krypsin, J. Organomet. Chem. **220**, 115 (1981); (c) A. Dedieu and A. Strich, Inorg. Chem. **18**, 2943 (1979); (d) A. Dedieu, Inorg. Chem. **20**, 2803 (1981)
28. (a) N. Koga, C. Daniel, J. Han, X.Y. Fu, and K. Morokuma, J. Am. Chem. Soc. **109**, 3455 (1987); (b) C. Daniel, N. Koga, J. Han, X.Y. Fu, and K. Morokuma, J. Am. Chem. Soc. **110**, 3773 (1988)
29. P. Cossee, J. Catal. **3**, 80 (1964)
30. O. Novaro, E. Blaisten-Barojas, E. Clementi, G. Giunchi, and M.E. Ruiz-Vizcaya, J. Chem. Phys. **68**, 2337 (1978). See also O. Novaro, Int. J. Quant. Chem. **42**, 1047 (1992)
31. H. Fujimoto, T. Yamasaki, H. Mizutani, and N. Koga, J. Am. Chem. Soc. **107**, 6157 (1985)
32. C.A. Jolly and D.S. Marynick, J. Am. Chem. Soc. **111**, 7968 (1989)
33. H. Kawamura-Kuribayashi, N. Koga, and K. Morokuma, J. Am. Chem. Soc. **114**, 2359 (1992)
34. H. Kawamura-Kuribayashi, N. Koga, and K. Morokuma, J. Am. Chem. Soc. **114**, 8687 (1992)
35. A. Anderson, H.G. Cordes, J. Herwig, W. Kaminsky, A. Merk, R. Mottweiler, J.H. Sinn, and H.-J. Vollmer, Angew. Chem., Int. Ed. Engl. **15**, 630 (1976)
36. S. Miya, T. Mise, and H. Yamazaki, Chem. Lett. 1853 (1989)
37. L.A. Castonguay and A.K. Rappe, J. Am. Chem. Soc. **114**, 5832 (1992)
38. E. Nakamura, Y. Miyachi, N. Koga, and K. Morokuma, J. Am. Chem. Soc. **114**, 6686 (1992)
39. (a) A.T. Stoll and E. Negishi, Tetrahedron Lett. **26**, 5761 (1985); (b) E. Nakamura, M. Isaka, and S. Matsuzawa, J. Am. Chem. Soc. **110**, 1297 (1988)
40. E. Nakamura, M. Nakamura, Y. Miyachi, N. Koga, and K. Morokuma, J. Am. Chem. Soc. **115**, 99 (1993)
41. W.F. Maier and P.v.R. Schleyer, **103**, 1891 (1981)

42. S. Nakamura and K. Morokuma, Abstracts, 33rd Symposium on Organometallic Chemistry, Japan; Tokyo, October 1986, Paper A109
43. L. Versluis and T. Ziegler, J. Am. Chem. Soc. **112**, 6763 (1990)
44. C. Bo and A. Dedieu, Inorg. Chem. 28, 304 (1989)
45. S. Sakaki and K. Ohkubo, Inorg. Chem. **27**, 2020 (1988); **28**, 2583 (1989)

NOBUAKI KOGA AND KEIJI MOROKUMA

CARBONYL MIGRATORY INSERTIONS

1. Introduction

Carbonyl insertion into an M-R bond, which is also called migratory insertion, (equation *1*) has been implicated as one of the key steps in various catalytic cycles such as hydroformylation [1]. When the migrating groups are alkyl, aryl, and alkenyl, insertion has been observed to take place in many transition metal complexes [1].

$$\text{(1)} \qquad \underset{\text{M—CO}}{\overset{\text{R}}{|}} \longrightarrow \text{M—}\overset{\text{O}}{\overset{\|}{\text{C}}}\text{—R}$$

There are two possible reaction pathways as shown in Scheme *1*: (i) migration of group R to carbonyl or (ii) literal CO insertion into the M-R bond.

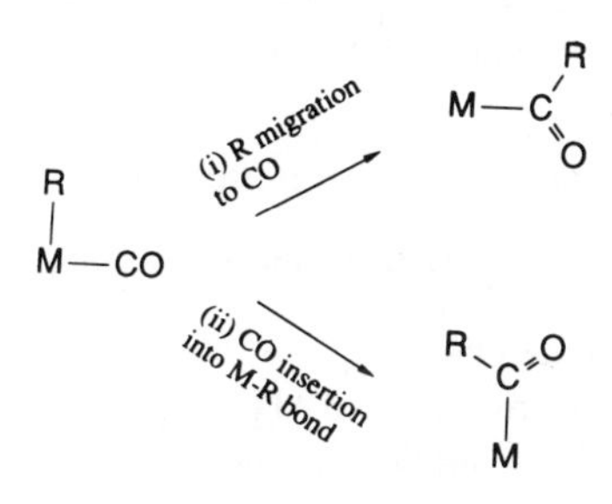

Scheme 1.

In order to address to this problem, many experimental studies have been carried out about the reactions for the complexes of various transition metals such as Mn [2], Rh [3], Ir [4], Fe [5], Pt [6], and Pd [7]. Due to the importance of this reaction in catalysis, the effects of the metal, ligands, and alkyl group substituents have also been investigated.

The hydride migratory insertion (R = H) has been known only for a few cases in transition metal complexes, [8] though it is believed to take place in

P.W.N.M. van Leeuwen et al. (eds.), Theoretical aspects of homogeneous catalysis, 93–113.

(2) $Cp^*Th(H)(OR) \xrightarrow{+CO} Cp^*Th(\eta^2\text{-}OCH)(OR)$

homogeneous catalytic reactions of carbon monoxide reductive hydrogenation [8ab]. In the actinide complex hydride migratory insertion can take place as in the example of reaction *2* with the activation free energy of 9 kcal/mol [9]. The exothermicity of 2 kcal/mol of this reaction has been ascribed to the oxophilicity of Th which favors the η^2-formyl complex.

The difficulty of hydride migratory insertion has been considered to be due to the strength of an M-H bond, which makes hydride migratory insertion endothermic [10]. Though the hydride migratory insertion may be difficult to study experimentally, it can be studied theoretically and compared with alkyl migratory insertion.

In order to discuss these problems, some theoretical studies have been carried out. We will show examples of such studies on carbonyl migratory insertion into M-CH_3 bonds in Section 2 and on that into M-H bonds in Section 3. In Section 4, the effect of substituents on the methyl group will be examined. Problems related to computational methods will be discussed in Section 5.

2. Carbonyl insertion into M-CH_3 bond

2.1. LATE TRANSITION METAL COMPLEXES: Pd, Pt, AND Co

In early MO studies, reactions of square planar complexes of group 10 transition metals have been investigated [11, 12]. Sakaki, Kitaura, Morokuma, and Ohkubo (SKMO) have studied reaction *3* at the restricted Hartree-Fock (RHF) level [11].

(3) $Pt(CH_3)(F)(CO)(PH_3)_2$ **1** $\rightarrow$ $Pt(COCH_3)(F)(PH_3)_2$ **2**

While the structures of the reactant and the product have been optimized, the geometry of the Pt complex has been changed stepwise to simulate three reaction paths, one for methyl migration, one for carbonyl migration and one for concerted migration of CO and methyl with simultaneous opening of the FPtP angle as shown in Scheme *2*. From the energy change along the assumed paths, it has been concluded that the easiest is methyl migration and that the most difficult is carbonyl migration.

Koga and Morokuma have studied the carbonyl insertion reaction *4* of d^8 square planar complexes [12].

(4) $M(CH_3)(H)(CO)(PH_3) \rightarrow M(COCH_3)(H)(PH_3)$
M = Pd **3** **5**
M = Pt **4** **6**

Their structure determination with the energy gradient method at the RHF level has shown that the transition state (TS) is three-centered and that the

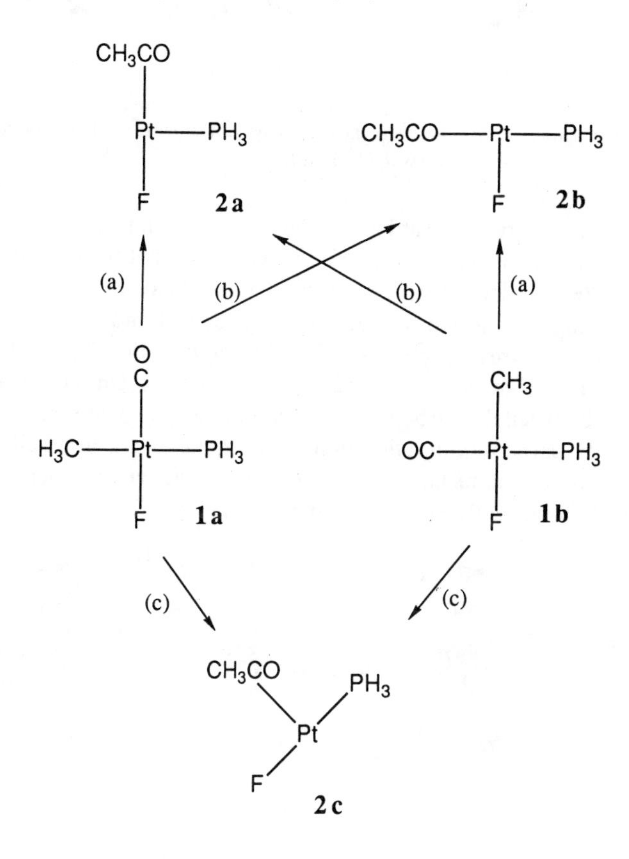

(a) CH_3 migration
(b) CO migration
(c) Concerted migration of CH_3 and CO

Scheme 2.

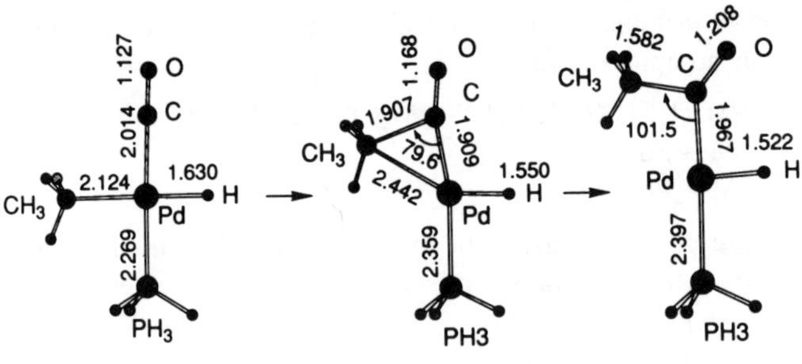

Reactant, 3 Transition State Product, 5

Fig. 1. RHF optimized structures (in Å and deg) of reactant, product, and transition state for carbonyl migratory insertion (reaction 4 with M = Pd) [12].

methyl group migrates toward the CO group as shown in Figure *1* for M = Pd.

Their analysis of the wave function has shown that the orbital interaction shown in **7a** stabilizes the three-centred TS, in which the sp^3 hybrid orbital of the migrating CH_3 group interacts with π^* of CO and the vacant d of the transition metal fragment. This is just in between the d – sp^3 bond in the reactant, **7b**, and the $sp^3 - \pi^*$ bond in the product, **7c**. On the other hand, the repulsion between the carbonyl lone pair and the occupied d orbital shown in **8** makes the carbonyl migration into the M-R bond less favorable.

Koga and Morokuma have also compared the reactivity in reaction *4* between Pd and Pt based on the metal-ligand bond energy [12]. As shown in Table *1*

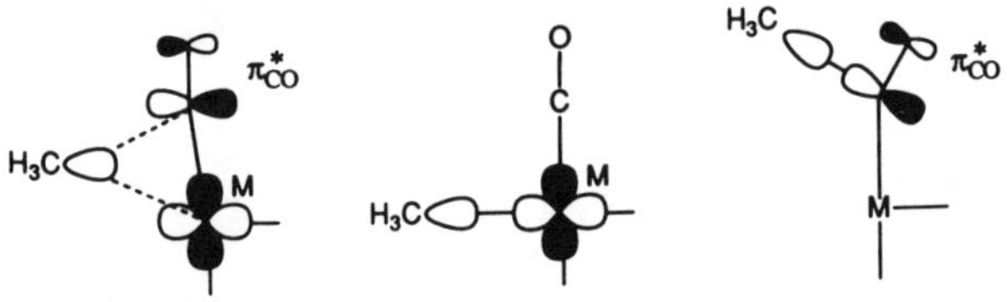

7a **7b** **7c**

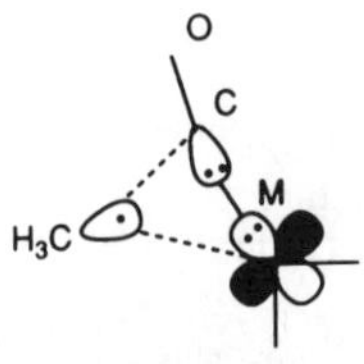

8

TABLE I

Energy of reaction (ΔE) and activation energy ($\Delta E^{\dagger}$) (in kcal/mol) of carbonyl insertion for (i) $Pd(CH_3)(H)(CO)(PH_3) \rightarrow Pd(COCH_3)(H)(PH_3)$ and (ii) $Pt(CH_3)(H)(CO)(PH_3) \rightarrow Pt(COCH_3)(H)(PH_3)$

	Method	$\Delta E^{\dagger}$	$\Delta E(\eta^1)$
i	RHF/I[b]	25.7	19.1
	RHF/II[b]	18.8	5.0
	MP2/II[b]	13.5	8.8
	RHF/III[c]	22.0	10.6
	MP2/III[c]	10.2	4.3
ii	RHF/I[b]	31.3	23.0
	RHF/II[b]	24.4	7.9
	MP2/II[b]	21.8	17.6

[a]The basis sets used consists of double-zeta valence basis functions with effective core potential for Pt and Pd and I) 3–21G for CH_3, CO, and hydride, and STO-2G for PH_3, II) [3s2p]/(9s5p) for C and O, [2s]/(4s) for all the hydrogens, and [6d4p]/(11s6p) for P, and III) [3s2p1d]/(9s5p1d) for C and O, [6d4s]/(11s6p) for P, and [2s]/(4s) for all the hydrogens. All the structures were optimized at the RHF/I level.

[b]Ref. 12.

[c]Ref. 21.

the Pt reaction is more endothermic and thus requires a larger activation energy. In the carbonyl insertion reaction, an M-R and an M-CO bond are broken, an M-COR and an R-CO bond are formed, and the CO triple bond changes into a double bond. Among these bonds, the bond energies of the first three metal-ligand bonds would depend on the central transition metal. Their analysis has shown that all the Pt-carbon bonds are stronger than the corresponding Pd-carbon bonds and consequently the Pt carbonyl insertion is less favorable, in which two strong Pt-carbon bonds are broken and only one strong Pt-carbon bond is formed. The stronger Pt-carbon bonds were ascribed to the larger relativistic effect in the heavier Pt atom. The relativistic effect stabilizes the Pt s orbital which is expected to participate in bonding more effectively than the Pd s orbital, thus making the Pt-carbon bond stronger.

Reaction 5 has been considered to be an elementary step of the hydroformylation catalytic cycle (R = alkyl) [13].

(5) $RCo(CO)_4 \rightarrow RCOCo(CO)_3$

Versluis, Ziegler, Baerends, and Ravenek (VZBR) have studied the model reaction with R = CH_3 by the Hartree-Fock-Slater (HFS) method [14]. They have optimized the structures of the reactant, $CH_3Co(CO)_4$, **9**, and the product, $CH_3COCo(CO)_3$, **10** and followed the path of reaction assuming that

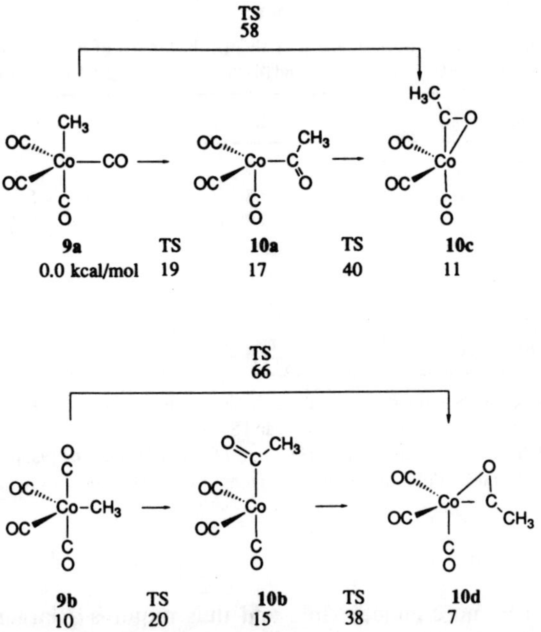

Fig. 2. Potential energy profile for carbonyl migratory insertion of $CH_3Co(CO)_4$, , in kcal/mol relative to **9a**, calculated with the HFS method. The transition states are assumed to be on the LST path [13].

TABLE II

Energy of reaction (ΔE) and activation energy ($\Delta E^{\dagger}$) (in kcal/mol) of carbonyl migratory insertion of $CH_3Mn(CO)_5$

Method	$\Delta E^{\dagger}$	$\Delta E(\eta^1)$	$\Delta E(\eta^2)$	Ref.
EH	20			15
HFS	21	18	− 1	16
PRDDO	18	1	−20	20
RHF[a]	–	3	–	20
RHF[b]	17	10	− 1	20

[a]The basis set used is double-ζ for 3s, 3p, 4s, and 4p and triple-ζ for 3d for Mn and the 4-31G for C, O, and H.

[b]The d polarization functions on C and O are added to the last basis set.

geometrical variable change linearly, i.e. the linear synchronous transit (LST) path, between the reactant and the product. The paths followed are for methyl migrations, **9a** → **10a** and **9b** → **10b** as well as CO insertions into the Co-

CH_3 bond, **9a** → **10c** and **9b** → **10d**. The results are summarized in Figure *2*. They have found that the activation energies for the methyl migrations are much smaller than for the CO insertions. The high activation energies were ascribed to the repulsion between the CO lone pair electrons and nonbonding d electrons, being the same as **8**. On the other hand, the products of the methyl migrations, **10a** and **10b**, with the η^1-coordination mode are less stable than those of the CO insertion 10c and 10d with the η^2-coordination mode. The additional coordination of the oxygen atom stabilizes **10c** and **10d**. Accordingly, the rearrangements from **10a** to **10c** and **10b** to **10d** take place through the acetyl group migrations with moderate activation energies.

2.2. MIDDLE TRANSITION METAL COMPLEX: Mn

The reaction mechanism, the substituent effects, and the energetics of the carbonyl migratory insertion of Mn complexes have been extensively studied experimentally [15]. Correspondingly the carbonyl insertion of an Mn complex (equation *6*) has been studied by several groups theoretically. The results are summarized in Table *II*.

11 → **TS** → η^1**-12** / η^2**-12** (6)

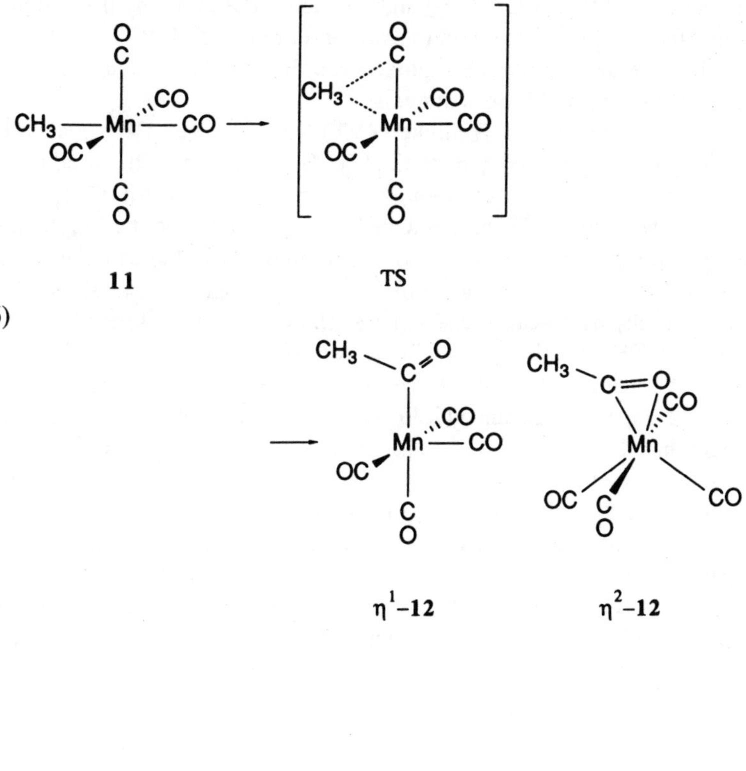

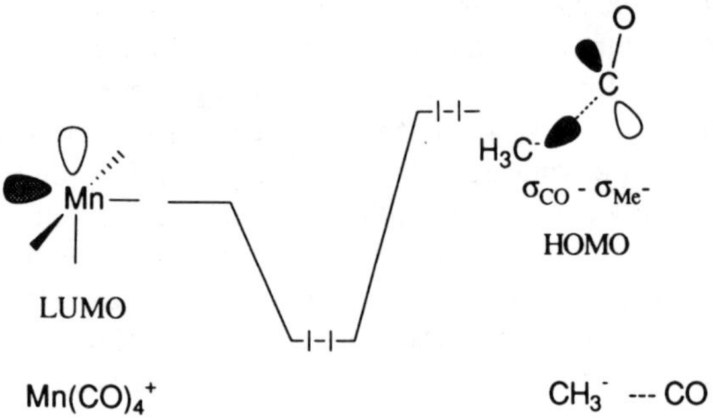

Mn
LUMO
$Mn(CO)_4^+$
H_3C
σ_{CO} - σ_{Me^-}
HOMO
CH_3^- --- CO

Scheme 3.

Berke and Hoffmann (BH) have searched with the extended Hückel (EH) method for the reaction path of the methyl migration which leads to η^1-**12** [16]. They discussed the low activation barrier of 20 kcal/mol in terms of the orbital interaction between CH_3CO^- and $Mn(CO)_4^+$ as shown in Scheme *3*.

In the uncatalyzed CH_3^- . . . CO system the σCO-σCH_3^- exchange repulsion raises the HOMO (σCO-σCH_3^-) high in energy, destabilizing the system. The interaction of this orbital with the vacant d orbital of $Mn(CO)_4^+$ results in the stabilization around the TS region. According to their calculation, the η^2-acetyl complex, η^2-**12**, does not exist.

Ziegler, Versluis, and Tschinke (ZVT) have followed BH's reaction path for reaction *6* by the HFS method [17]. The structure of the product acetyl ligand has been optimized, though the structure of the $Mn(CO)_4$ fragment has been taken from the experiments. They have obtained the endothermicity of 18 kcal/mol for the formation of the η^1 product (η^1-**12**) with the activation barrier of 21 kcal/mol as shown in Table *II*. Although the activation barrier is different only by 1 kcal/mol, Ziegler *et al.* have found, different from the EH results by BH, that the η^2-acetyl complex (η^2-**12**) is more stable than the η^1-**12** by 19 kcal/mol. As a result, the overall reaction is 1 kcal/mol exothermic. In η^2-**12** the Mn-O distance (2.30 Å) has been calculated to be substantially longer than the Mn-C bond (1.90 Å), suggesting relatively weak η^2-interaction. This weak interaction, where electrostatic attraction plays an important role, could not be described by the EH method. The longer M-O bond is different from the bond found for early transition metal complexes which have been studied both theoretically [18, 19] and experimentally [20]. In the early transition metal complex, the M-O interaction would be stronger because of its electron deficiency as will be shown in the reaction of a Sc formyl complex.

Axe and Marynick (AM) have studied the same methyl migration (equation *6*), [21] by determining the TS structure at the approximate *ab initio* level with the partial-retention-of-diatomic-differential- overlap (PRDDO) approximation and calculated the energetics at the full *ab initio* RHF level. They have shown that methyl migration takes place with a moderate activation barrier and that the orbital interaction shown in **7a** is important. Also, they have found that the η^2-acyl complex is more stable than the η^1-acyl complex, a similar conclusion to that by ZVT shown above [16].

TABLE III

Energy of reaction (ΔE) and activation energy ($\Delta E^\dagger$) (in kcal/mol) of carbonyl insertion for $Pd(H)_2(CO)(PH_3) \rightarrow Pd(CHO)(H)(PH_3)$[a]

Method	$\Delta E^\dagger$	$\Delta E(\eta^1)$
RHF/I	26.9	25.4
RHF/III	19.2	17.7
MP2/III	8.2	9.4

[a]The basis sets used consists of [2s2p2d]/(3s3p4d) with effective core potential for Pd, 3-21G for CH_3, CO, and hydride, and STO-2G for PH_3 and III) [3s2p1d]/(9s5p1d) for C and O, [6d4s]/11s6p) for P, [3s1p]/(5s1p) for the hydride and [2s]/(4s) for the other hydrogens. All the structures were optimized at the RHF/I level. Ref. 21.

Concluding the section, theoretical calculations have shown that in late and middle transition metal complexes with d lone pair electrons the alkyl migration is more favorable than the CO insertion into the M-R bond. The latter reaction path is less favorable because of the repulsion between CO lone pair electrons and lone pair d electrons. The activation energy for methyl migration was calculated to be modest.

3. Carbonyl migratory insertion into M-H bond

3.1. LATE TRANSITION METAL COMPLEX: Pd AND Co

Koga and Morokuma have adopted the hydride migration reaction *7* to compare the reactivity directly with methyl migration 4, where the ligand *trans* to the migrating group is the same in both reactions [22]. The energetics is shown in Table *III*. Hydride migration 7 is slightly more endothermic than methyl migration 4 at the MP2 level. This larger endothermicity of reaction *7* has been ascribed to the fact that the Pd-H bond is stronger than the Pd-CH_3 bond by 5 kcal/mol. At the MP2 level, the barrier for the reverse

(7) H—Pd(CO)(PH$_3$)—H → TS → (HCO)Pd(PH$_3$)—H

TS

decarbonylation becomes negligible. In fact the geometry optimization with the MP2 energy gradient has shown that the potential energy surface is downhill for decarbonylation, indicating that the η^1-formyl complex is kinetically unstable. The three-centered interaction at the three-centered 'TS' is stabilized by the orbital interaction similar to **7a**, in which the spherical 1s orbital could strongly interact simultaneously with the CO π^* orbital and the spd hybrid orbital of the transition metal. This strong interaction because of the non-directionality of the hydride 1s orbital would make the transition state disappear. The fact that the hydride migration is up-hill is probably the reason why the formyl complex is not detected. However, the endothermicity of hydride migration is low enough for the formyl complex to be an important transient species in catalytic processes. Koga and Morokuma have also found that, although the formyl complex is not stationary, it is on the reaction path to $(H_2CO)Pd(PH_3)$, dual hydride migration to CO.

Reaction *8* should be compared with model reaction *5* of the elementary step in hydroformylation catalytic cycle shown in Section 2.1.

(8) $HCo(CO)_4$, **13** → $HCOCo(CO)_3$, **14**

While Antolovic and Davidson (AD) [23] have treated this reaction as a model reaction of the methyl migration, Versluis, Ziegler, Baerends, and Ravenek (VZBR) [13] have studied this for comparison with the methyl migration (equation *5*).

In the HFS calculations by VZBR the structures of **13** and **14** have been optimized and the reaction path has been followed by the LST approximation, the same method as used for the methyl migration mentioned above [13]. The reaction steps they have studied are shown in Figure *3*.

It has been found that neither **14a** nor **14b** (kinetic product) exists. The HFS geometry optimizations of these species with the C_s symmetry constraint have led to a hydride carbonyl complex, indicating that these formyl complexes are kinetically unstable. The equilibrium structures of **14**'s they have

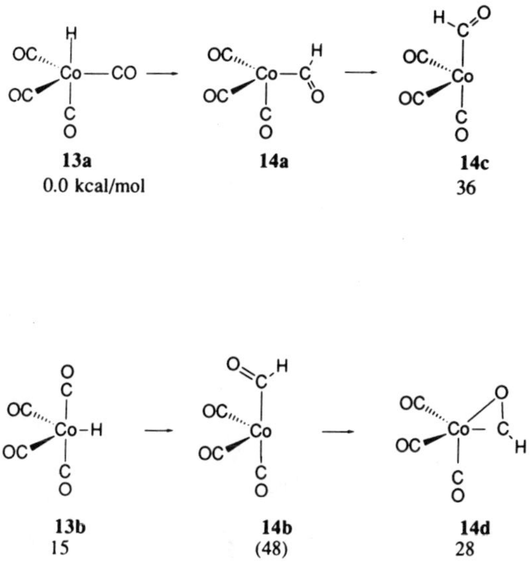

13a 0.0 kcal/mol — **14a** — **14c** 36

13b 15 — **14b** (48) — **14d** 28

Fig. 3. Potential energy profile for carbonyl migratory insertion of $HCo(CO)_4$, **13**, in kcal/mol relative to **13a**, calculated with the HFS method. The transition states are assumed to be on the LST path [13].

determined, **14c** and **14d**, have a vacant coordination site anti to the (migrating) hydrogen of the formyl group, and thus are not kinetic products. The kinetic products, **14a** and **14b**, could transform into more stable isomers through formyl migration **14a** → **14c** or **14b** → **14d** or Co-CHO rotation **14a** → **14d** or **14b** → **14c** (the authors have not discussed the latter process). **14d** is more stable than **14c**, since in **14d** the η^2 interaction takes place. The **13a** → **14c** and **13b** → **14d** processes are 36 and 13 kcal/mol endothermic, respectively, which are more endothermic than the corresponding methyl migration in Figure *2*. The higher endothermicity has been ascribed to the Co-H bond which is stronger than the Co-CH_3 bond.

AD have determined the structures of $HCo(CO)_4$, **13**, and $HCOCo(CO)_3$, **14**, by the RHF energy gradient method [22] and have further followed the reaction path using the LST approximation. They have found an activation energy larger than 50 kcal/mol, which is too high for a catalytic reaction. Probably, responsible to this from the computational aspect are the lack of electron correlation and the unoptimized TS structure. In addition, there are some problems in the RHF calculation for the Co carbonyl complex [24, 25]; the most stable RHF structure of **13** is trigonal bipyramid (TBP) with equatorial

(9)

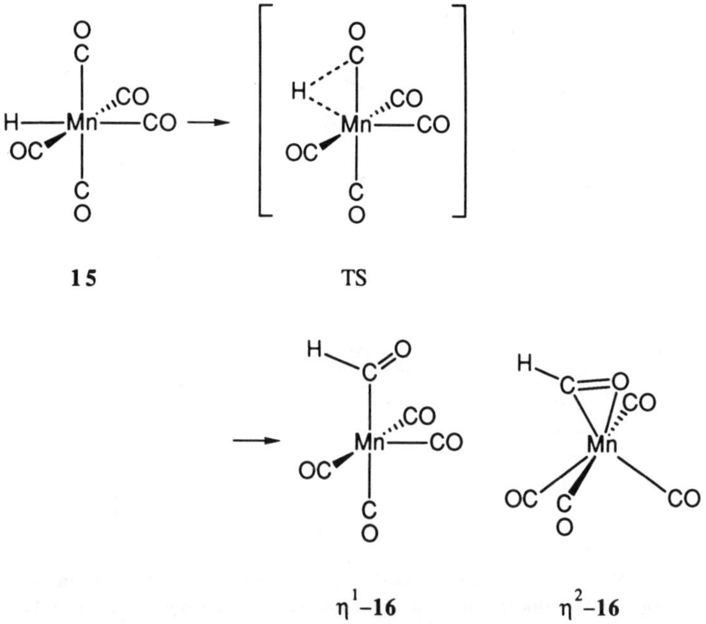

15 TS

η^1-16 η^2-16

H, while the experimental structures are TBP with apical H. One must take into account the electron correlation effect properly. It is effectively taken into account in the calculations by VZBR through the HFS correction term. Therefore, the results by AD are different from those of VZBR mentioned above and are not conclusive. However, they have ascribed the difficulty of H migration to the strong M-H bond. On the other hand, the HFS results seem to be more reasonable than the RHF results.

3.2. MIDDLE TRANSITION METAL COMPLEX: Mn

The CO insertion into the Mn-H bond (equation *9*) has been investigated as model reaction of reaction *6* or for comparison in the studies of reaction *6*. The results are summarized in Table *IV*.

Berke and Hoffmann have calculated [15] the hydride migration with the EH method to obtain an activation energy of 16 kcal/mol, 4 kcal/mol smaller than that of the methyl migration (see Section 2.2). This lower barrier has been ascribed to the stabilization of the orbital which develops into the new C-H bond. Note that this origin is different from those proposed with the

TABLE IV

Energy of reaction (ΔE) and activation energy ($\Delta E^{\dagger}$) (in kcal/mol) of carbonyl migratory insertion of $HMn(CO)_5$

Method	$\Delta E^{\dagger}$	$\Delta E(\eta^1)$	$\Delta E(\eta^2)$	Ref.
EH	16			15
RHF[a]	14	11	–	25a
SDCI[a]	39	38	–	25a
HFS	40	38	22	16
PRDDO	20	14	–5	26
RHF[b]	–	18	–	26
RHF[c]	–	26	15	26

[a]The basis functions used are [5s3p3d]/(13s8p6d) for Mn, [3s2p]/(9s5p) for C and O, and [3s]/(6s).
[b]The basis set used is double-ζ for 3s, 3p, 4s, and 4p and triple-ζ for 3d for Mn and the 4–31G for C, O, and H.
[c]The d polarization functions on C and O are added to the last basis set.

more sophisticated methods discussed below. Several other aspects of migratory insertion have been discussed as well in this EH studies. Although the conclusions obtained can explain qualitatively what is happening and are consistent with those of the following more accurate *ab initio* MO and HFS studies, the hydride migration 9 is very endothermic and requires a larger activation energy as discussed below.

Nakamura and Dedieu (ND) have studied reaction *9* up to η^1-**16** [26]. They have determined at the RHF level the reaction path by partial geometry optimization without using the energy gradient technique, finding that the mechanism is hydride migration. The orbital interaction schemes similar to **7a** and **8** were given to rationalize the hydride migration [25b]. Along this reaction path, they have carried out more reliable single-and-double excitation configuration interaction (SDCI) calculations to include the electron correlation effect. They have found that the barrier is 39 kcal/mol and that the endothermicity is 38 kcal/mol, suggesting that hydride migration is quite difficult and even if the hydride migrates, the reverse reaction would take place with a very small barrier.

Ziegler, Versluis, and Tschinke (ZVT) have carried out an HFS study on reaction *9*, with the same procedure as in their study of the methyl migration discussed in Section 2.2. [16]. ZVT have obtained the endothermicity of 38 kcal/mol for the formation of the η^1 product (η^1-**16**) with the activation barrier of 40 kcal/mol as shown in Table *IV*. It is interesting to see that this endothermicity coincides with that calculated at the SDCI level by ND [25a], noting that the HFS method in part takes the electron correlation into account.

The hydride migration is much more endothermic than the methyl migration which has a calculated endothermicity of 18 kcal/mol. The difference in endothermicity between the two reactions was ascribed to the difference in bond strength between Mn-CH_3 and Mn-H bonds: the latter has been calculated to be 12 kcal/mol stronger than the former with the HFS method. This is in good agreement with the experimental difference of 14 kcal/mol [10b]. ZVT have found that η^2-formyl complex is more stable than η^1-formyl complex, similar to the acetyl complex discussed in 2.2.

AM have also compared methyl and hydride migration of the Mn complex using the PRDDO and the *ab initio* RHF method, to obtain the results similar to those by ZVT [27]. The more endothermic hydride migration is again ascribed to the stronger Mn-H bond than the Mn-CH_3 bond; the former is calculated by AM to be stronger by about 20 kcal/mol. They also found the η^2-formyl complex is more stable than the η^1-formyl complex.

One could conclude based on the above theoretical studies that the hydride migration of $HMn(CO)_5$ is very endothermic and that the reverse reaction can take place with a quite small activation energy. The hydride migration of the Mn complex is more endothermic than that of the Pd complex discussed in Section 3.1. The back-donation in the Mn(I) complex is stronger than in the Pd(II) complex as will be discussed in Section 5. The change in strong π back-donation during the reaction, i.e., from out-of-plane and in-plane π back-donations in the reactant to only in-plane π back-donation in the product, is essentially responsible to this large endothermicity. Berke and Hoffmann have also suggested that in the complex of heavier metals with more diffuse and less stable d orbitals, the π back-donation would make a large contribution to the M-CO bonding and thus the loss of in-plane π back-donation would affect the energetics [15].

3.3. EARLY TRANSITION METAL COMPLEX: Sc

Rappé has studied hydride migratory insertion in an early transition metal complex (equation *10*) [28]. This model reaction was chosen because of its simplicity.

$$(10) \qquad Cl_2ScH + CO \rightarrow Cl_2Sc(CHO)$$

The structures of the stationary points have been determined by the RHF energy gradient method and the energy calculations were carried out at the generalized valence bond (GVB) and CI level of calculation. The resultant potential energy profile is shown in Figure *4*.

For the formyl complex, only the η^2 structure exists, which is more stable

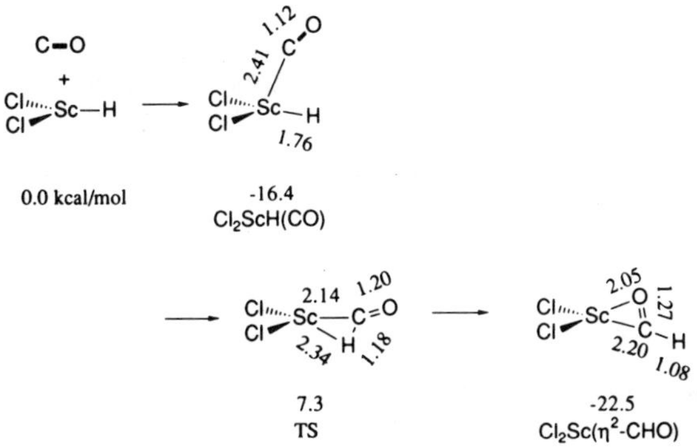

0.0 kcal/mol -16.4 $Cl_2ScH(CO)$

7.3 TS -22.5 $Cl_2Sc(\eta^2\text{-}CHO)$

Fig. 4. Potential energy profile for CO insertion, $Cl_2ScH + CO$ in kcal/mol calculated at the GVB-CI level, relative to the reactants. The selected bond lengths are shown in Å [27].

than the reactant complex, $Cl_2ScH(CO)$. This is clearly in contrast to the Mn formyl complex shown in Section 3.2. Scandium is electron-deficient and thus the bonding interaction between the lone pair of formyl oxygen and the empty d orbital of Sc has been found to favor the η^2 coordination. This is in agreement with the previous extended Hückel studies of an η^2-acyl complex [29] and the experimental observation of reaction *2*.

The conclusion of this section is that carbonyl migratory insertion into an M-H bond proceeds in the middle and late transition metal complexes through hydride migration, similar to the methyl migration mechanism in the insertion into an M-C bond. Compared with the corresponding methyl migrations, the hydride migrations which give the η^1-formyl complexes are more endothermic and the activation energies of the reverse decarbonylations are small. On the other hand, the η^2-coordination of the formyl ligand stabilizes the product, specially in the early transition metal complex.

4. Effect of substituents and π^* orbital energy on carbonyl migratory insertions

The effect of substituent on the migrating alkyl group has been studied by Koga and Morokuma (KM) in reaction *4* with M = Pd [12] and Axe and Marynick (AM) in reaction *6* [26] as shown in Table *V*. KM have discussed the effect of substituents in terms of the M-R bond energies. KM have stated that electronegative substituents stabilize the alkyl ligand that has a negative charge, leading to the stronger M-R bond as shown in the larger endothermicity and the

TABLE V

Effects of migrating group on energy (ΔE) and activation energy ($\Delta E^{\dagger}$) (in kcal/mol) of carbonyl insertions. The reactants are $Pd(R)(CO)(H)(PH_3)$ and $RMn(CO)_5$

R	$\Delta E^{\dagger}$	$\Delta E(\eta^1)$	$\Delta E(\eta^2)$
$Pd(R)(CO)(H)(PH_3)$[a]			
CH_2CH_3	23	15	–
CH_3	26	19	–
CHF_2	41	31	–
$RMn(CO)_5$[b]			
CH_2CH_3	–	–8	–
CH_3	–	3(10)	(–1)
CF_3		26(28)	(31)

[a]Calculated at the RHF level. Ref. 12.
[b]Calculated at the RHF level. Numbers in parentheses with the PRDDO approximation. Ref. 26.

higher activation energy for difluoromethyl. AM's results show a similar trend.

AM have compared for the reaction *6* the orbital interaction between the migrating group orbital and the metal spd hybrid orbital in the reactant (**7b**) with that between the migrating group orbital and the CO π^* orbital in the product (**7c**) [26]. The spd hybrid is more stable than CO π^*. If the π^* orbital is more stable, for instance like in CS, or the spd hybrid is less stable, for instance like in Co, the interaction in the product is strengthened, resulting in the lower energy requirement. ZVT [16] have also proposed a same consideration of the π^* orbital energy in their studies of the reactions *6* and *9* and have compared the hydride migration to CS (equation *11*) with that to CO (equation *9*) by the HFS method. Experimentally, it has been found that the migration to a CS ligand is preferred to the migration to a CO ligand [30]. Even the hydride has been observed to migrate to CS in an Os complex [31].

$$(11) \quad \underset{\mathbf{17}}{HMn(CO)_4(CS)} \rightarrow \underset{\mathbf{18}}{HCSMn(CO)_4}$$

Reaction *11* giving the η^1-thioformyl complex (η^1-**18**) has been calculated to be 17 kcal/mol endothermic, which is compared with the endothermicity of 38 kcal/mol for the reaction *9*. Since the Mn-H bond is broken in both reactions *9* and *11*, the difference is expected to originate from the difference in the H-CX bond strength (X = O or S). Their calculations have shown that the $H\text{-}CSMn(CO)_4$ bond is 20kcal/mol stronger than the $H\text{-}COMn(CO)_4$ bond, consistent with the stronger interaction of H 1s with CS π^* discussed above. Calculations have also shown that the rearrangement from η^1-**18** to η^2-**18** is 31 kcal/mol exothermic and thus the formation of η^2-**18** from **17** is 14 kcal/mol exothermic overall, supporting the experimental suggestion of a stable η^2-thioformyl complex [30c].

5. Remarks on electron correlation and basis set effects

The calculations of reaction *9* by Nakamura and Dedieu have shown that electron correlation increases the activation barrier by 25 kcal/mol and the endothermicity by 27 kcal/mol [25]. The large energetics difference between SCF and CI potential energy surfaces was ascribed to the unbalanced description of π back-donation between the reactant Mn-CO bond and the product Mn-CHO bond at the SCF level. Dedieu, Sakaki, Strich, and Siegbahn (DSSS) have carried out an *ab initio* complete active space (CAS) SCF and MP2 calculation for reaction *12* with an assumed structure in order to investigate this difference in more detail (Table *VI*) [32]. The electron correlation effects increase exothermicity in reaction *12* as in reaction *9*.

(12) $\quad RMn(CO) \rightarrow Mn(RCO)$ $R = H, CH_3$

Analysis of the CASSCF wave function has demonstrated that the correct description of π back-donation in the Mn complexes requires a multi-determinantal wave function such as in CI, CASSCF, Møller-Plesset perturbation theory. The Hartree-Fock method underestimates the π back-donation in the Mn complexes. Though both in-plane and out-of-plane π back-donations take place in the reactant, only out-of-plane π back-donation can take place in the product, since the in-plane CO π^* orbital changes to the C-H σ bond orbital during the reaction. In addition, the formyl ligand in the product is negatively charged and thus back-donation is weak. This crucial difference in π back-donation between the reactant and the product cannot be described by the HF method. The HF level of calculations artificially underestimates the π back-donation in the reactant to favor the product.

DSSS have also carried out an *ab initio* calculation for a Pd reaction (equation *13*) as shown in Table *VI*, using an assumed structure [31].

(13) $\quad HPd(CO)^+ \rightarrow Pd(CHO)^+$

This was chosen as a model reaction of reaction *4*, in which the correlation effect is smaller than that in Mn reaction *9*. The electron correlation effect on the reaction *13* makes the reaction more exothermic, as in reaction *4*. DSSS

TABLE VI

Energy of reaction (in kcal/mol) of carbonyl insertion at several levels of calculation

Reactant	SCF	CASSCF4	CASSCF10	MP2	SDCI	Ref.
HMn(CO)	2.3[a]	15.1	35.5	35.5	23.3	31
	11.5[b]	–	–	–	27.4	33
	13.9[c]	–	–	–	–	33
	7.4–15.2[d]	–	–	–	–	32
	19.5–25.4[e]	–	–	–	–	32
$CH_3Mn(CO)$	−1.8[a]	–	–	36.4	–	31
$HPd(CO)^+$	−10.2[a]	−21.4	−33.6	−15.8	–	31

[a]The basis functions used are [5s3p3d]/(13s8p6d) for Mn, [6s4p5d]/(15s9p8d] for Pd, [3s2p]/(9s5p) for C and O, and [3s]/(6s) (basis set I).
[b]One set of p and d polarization functions on H, C and O are added to the basis set I.
[c]Two sets of p and d polarization functions on H, C and O are added to the basis set I.
[d]The basis sets used do not include polarization functions.
[e]The basis sets used include polarization functions.

have found that the electron correlation effect on Pd-H and Pd-CHO σ bonds is essential and that the π back-donation is not as important in the Pd complex as in the Mn complex, because the d orbital of Pd(II) is more stable than that of Mn(I) due to the higher oxidation number. DSSS have concluded that the correlation effect may vary according to the nature and oxidation state of the transition metal atom.

Axe and Marynick have carried out the RHF calculations for Mn hydride migration (equation *12*) with several larger polarized basis sets, in order to investigate the effects of the polarization functions [33]. The RHF exothermicity calculated with unpolarized basis sets is between 9 and 15 kcal/mol, while that calculated with polarized basis sets is between 20 and 25 kcal/mol. The latter value is closer to the SDCI value calculated with an unpolarized basis set [31]. Polarization functions seems to improve the π back-donation as well. Therefore, the electron correlation effect found by DSSS might be overestimated. Dedieu has later calculated the energy of Mn hydride migration (equation *12*) by the SDCI method with a polarized basis set as shown in Table *VI* [34]. The electron correlation effect calculated with an unpolarized basis set has been found actually to be overestimated. However, the electron correlation effect still amounts to 16 kcal/mol with the polarized basis set. Both polarization functions and the correlation effect are indispensable for a quantitative calculation of the Mn reaction through a correct description of π-back donation.

6. Summary

In this chapter, we reviewed theoretical studies on carbonyl migratory insertion into an M-R (R = CH_3 and H) bond. These showed that in the late and middle transition metal complexes the methyl or hydride migrates to CO and that the CO group does not insert into the M-R bond, which is unfavorable because of the repulsion between CO lone pair and lone pair d electrons. As expected, the hydride migration is more endothermic than the methyl migration, since the M-H bond is stronger.

Several reactions closely related have been studied theoretically. Some carbonyl migratory insertions into the M-H bond on a transition metal atom without any other ligands have been studied: the reactions for HRh(CO) [35] and $HM(CO)^n$ (M = Rh,Pd; n = 0, +1) [36] They are intended to be models of the reaction on a surface and thus are not directed to the homogeneous catalyst. In the CH_2 ligand, there is a vacant p_π orbital which can play the same role as the CO π^*. Reaction, $ClRu(CH_2)(H) \rightarrow ClRuCH_3$ has been studied; the TS is three-centered, similar to that for hydride migration to CO and the activation barrier of 12 kcal/mol is low [37].

Nobuaki Koga
School of Informatics and Sciences and Graduate School of Human Informatics
Nagoya University, Nagoya 464-01, Japan

Keiji Morokuma
Cherry L. Emerson Center for Scientific Computation and Dept. of Chemistry
Emory University, Atlanta, GA 30323, USA

References

1. (a) J.P. Collman, L.S. Hegedus, J.R. Norton and R.G. Finke, *Principles and Applications of Organotransition Metal Chemistry*; University Science Books: Mill Valley, California, (1987); (b) F.A. Cotton and G. Wilkinson, *Advanced Inorganic Chemistry*; Wiley: New York, 1988. (c) A. Yamamoto, Organotransition Metal Chemistry; Wiley: New York (1986)
2. (a) K. Noack and F. Calderazzo, J. Organomet. Chem. **10**, 101 (1967); (b) K. Noack, M. Ruch and F. Calderazzo, Inorg. Chem. **7**, 345 (1968); (c) T.C. Flood, J.E. Jensen and J.A. Statler, J. Am. Chem. Soc. **103**, 4410 (1981)
3. (a) D.A. Slack, D.L. Egglestone and M.C. Baird, J. Organomet. Chem. **146**, 71 (1978); (b) D.L. Egglestone, M.C. Baird, C.J.L. Lock and G. Turner, J. Chem. Soc. Dalton Trans. 1576 (1979)
4. R.W. Glyde and R.J. Mawby, Inorg Chim. Acta **5**, 317 (1971)
5. (a) T.G. Attig and A. Wojcicki, J. Organomet. Chem. **82**, 397 (1974); (b) P. Reich-Rohrwig and A. Wojcicki, Inorg. Chem. **13**, 2457 (1974); (c) A. Davison and N. Martinez, J. Organomet.

Chem. **74**, C17 (1974); (d) T.C. Flood and K.D. Campbell, J. Am. Chem. Soc. **106**, 2853 (1984); (e) T.C. Flood, K.D. Campbell, H.H. Downs and S. Nakanishi, Organometallics **2**, 1590 (1983); (f) H. Brunner, B. Hammer, I. Bernal and M. Draux, Organometallics **2**, 1595 (1983); (g) S.C. Wright and M.C. Baird, J. Am. Chem. Soc. **107**, 6899 (1985)

6. G.K. Anderson and R.J. Cross, J. Chem. Soc. Dalton Trans. 1246 (1979)
7. F. Ozawa and A. Yamamoto, Chem. Lett. 289 (1981)
8. (a) Herrmann, W.A., Angew. Chem. Int. Ed. Engl. **21**, 117 (1982); (b) J.R. Blackborow, R.J. Daroda and G. Wilkinson, Coord. Chem. Rev. **43**, 17 (1982); (c) D.L. Grimett, J.A. Labinger, J.N. Bonfiglio, S.T. Masuo, E. Shearin and J.S. Miller, Organometallics **2**, 1325 (1983); (d) C. Masters, Adv. Organomet. Chem. **17**, 61 (1979); (e) D.R. Fahey, J. Am. Chem. Soc. **103**, 136 (1981)
9. K.G. Moloy and T.J. Marks, J. Am. Chem. Soc. **106**, 7051 (1984)
10. (a) Ref. 1a p.369; (b) J.A. Connor, M.T. Zafarani-Moattar, J. Bickerton, N.I. El Saied, S. Suradi, R. Carson, G. Al Takhin and H.A. Skinner, Organomet allics **1**, 1166 (1982); (c) K.L. Lane, L. Sallans and R.R. Squires, Organometallics **4**, 408 (1984).
11. S. Sakaki, K. Kitaura, K. Morokuma and K. Ohkubo, J. Am. Chem. Soc. **105**, 2280 (1983)
12. N. Koga and K. Morokuma, J. Am. Chem. Soc. **107**, 7230 (1985); **108**, 6536 (1986)
13. R.F. Beck and D.S. Breslow, J. Am. Chem. Soc. **83**, 4023 (1961)
14. L. Versluis, T. Ziegler, E.J. Baerends and W. Ravenek, J. Am. Chem. Soc. **111**, 2018 (1989)
15. (a) T.H. Coffield, J. Kozikowski, R.D. Closson, J. Org. Chem. **22**, 598 (1957); (b) F. Calderazzo and F.A. Cotton, Inorg. Chem. **1**, 30 (1962); (c) K.A. Keblys and A.H. Filbey, J. Am. Chem. Soc. **82**, 4204 (1960); (d) R.J. Mawby, F. Basolo and G. Pearson, J. Am. Chem. Soc. **86**, 3994 (1964); (e) F. Calderazzo and K. Noack, K. Coord. Chem. Rev. **1**, 118 (1966); (f) K. Noack and F. Calderazzo, J. Organomet. Chem. **10**, 101 (1967); (g) J.N. Cawse, R.A. Fiato, R.L. Pruett, J. Organomet. Chem. **172**, 405 (1979); (h) T.M. McHugh, A.J. Rest, J. Chem. Soc., Dalton Trans. 2323 (1980); (i) T.C. Flood, J.E. Jensen and J.A. Statler, J. Am. Chem. Soc. **103**, 4410 (1981)
16. H. Berke and R. Hoffmann, R. J. Am. Chem. Soc. **100**, 7224 (1978)
17. T. Ziegler, L. Versluis and V. Tschinke, J. Am. Chem. Soc. **108**, 612 (1986)
18. A.K. Rapp, J. Am. Chem. Soc. **109**, 5605 (1987)
19. K. Tatsumi, A. Nakamura, P. Hofmann, P. Stauffert and R. Hoffmann, J. Am. Chem. Soc. **107**, 4440 (1985)
20. (a) G. Fachinetti, G. Fochi and C. Floriani, J. Chem. Soc., Dalton Trans. 1946 (1977); (b) K.G. Moloy and T.J. Marks, J. Am. Chem. Soc. **106**, 7051 (1984) and references cited therein
21. F.U. Axe and D.S. Marynick, Organometallics **6**, 572 (1987)
22. N. Koga and K. Morokuma, New. J. Chem. **15**, 749 (1991)
23. (a) D. Antolovic and E.R. Davidson, J. Am. Chem. Soc. **1987**, **109**, 5828 (1987); (b) D. Antolovic and E.R. Davidson, J. Am. Chem. Soc. **109**, 977 (1987)
24. D. Antlovic and E.R. Davidson, J. Chem. Phys. **88**, 4967 (1988)
25. H.P. Luthi, P.E.M. Siegbahn and J. Almlof, J. Phys. Chem. **89**, 2156 (1985)
26. (a) S. Nakamura and A. Dedieu, Chem. Phys. Lett. **111**, 243 (1984); (b) A. Dedieu and S. Nakamura, in *Quantum* Chemistry: The Challenge of Transition Metals and Coordination Chemistry, Veillard, A., Ed., NATO ASI series, Vol 176; Reidel: Dordrecht, 1986, pp. 277
27. F.U. Axe and D.S. Marynick, J. Am. Chem. Soc. **110**, 3728 (1988)
28. A.K. Rappe, J. Am. Chem. Soc. **109**, 5605 (1987)
29. M.D. Curtis, K.-B. Shiu and W.M. Butler, J. Am. Chem. Soc. **108**, 1550 (1986)
30. G.B. Clark, T.J. Collins, K. Marsden and W.R. Roper, J. Organomet. Chem. **259**, 215 (1983)
31. (a) T.J. Collins and W.R. Roper, J. Chem. Soc., Chem. Commun. 104 (1976); (b) W.R. Roper

and K.G. Town, J. Chem. Soc., Chem. Commun. 781 (1977); (c) T.J. Collins and W.R. Roper, J. Organomet. Chem. **73**, 159 (1978)
32. A. Dedieu, S. Sakaki, A. Strich and P.E.M. Siegbahn, Chem. Phys. Lett. **133**, 317 (1987)
33. F.U. Axe and D.S. Marynick, Chem. Phys. Lett. **141**, 455 (1987)
34. A. Dedieu, in The Challenge of d and f Electrons Theory and Computation, Salahub, D.R. and Zerner, M.C., Eds., American Chemical Society symposium series, No. 394; American Chemical Society: Washigton, DC, 1989, pp. 58
35. M.L. McKee, C.H. Dai and S.D. Worley, J. Phys. Chem. **92**, 1056 (1988)
36. G. Pacchioni, P. Fantucci, J. Koutecky and V. Ponec, J. Catal. **112**, 34 (1988)
37. E.A. Carter and W.A. Goddard, J. Am. Chem. Soc. **109**, 579 (1987)

ELZBIETA FOLGA, TOM WOO AND TOM ZIEGLER

A DENSITY FUNCTIONAL STUDY ON $[2_s + 2_s]$ ADDITION REACTIONS IN ORGANOMETALLIC CHEMISTRY

Density Functional (DF) calculations have been carried out on organometallic $[2_s + 2_s]$ addition processes. The study includes the σ-bond metathesis reaction (I): $Cp_2Sc\text{-}R + H\text{-}R' \rightarrow Cp_2Sc\text{-}R' + H\text{-}R$ ($R = H, CH_3$; $R' = H, CH_3$, vinyl and acetylide); the double bond metathesis reaction (II): $Cl_2Mo(O)CH_2 + H_2C = CH_2 \rightarrow Cl_2Mo(O)CH_2 + H_2C = CH_2$; and the triple bond exchange reaction (III): $Cl_2MoCH + HC \equiv CH \rightarrow Cl_2MoCH + HC \equiv CH$. An energy profile has been traced for each of the three reaction types with characterization of the four-center transition states (I) or intermediates (II, III) involved. Information is further provided on the geometries and relative energies of all reactants and products. The organometallic $[2_s + 2_s]$ addition reactions are compared to the corresponding organic processes in an analysis based on qualitative molecular orbital theory.

1. Introduction

The $[2_s + 2_s]$ addition reaction of equation *1*

(1) A–M + C–D → [A–C, M–D four-center] **1**; Eq. 1a → A–C + M–D; Eq. 1b → A–M, C–D, M–D

is now recognized as essential in many organometallic processes of importance for catalysis. The pivotal step in the addition reaction is the formation of a four center transition state or intermediate, **1**, containing at least one metal atom.

The reaction in equation *1b* was recognized early on [1] as a key step in Ziegler-Natta polymerization of olefins (and acetylenes):

(2) L_nM–C(H)(H)(P) + olefin–R → [L_nM ··· C(H)(H)(P) ··· C–R, C] **2** → L_nM–C–C(R)–C(H)(H)(P)

P.W.N.M. van Leeuwen et al. (eds.), Theoretical aspects of homogeneous catalysis, 115–165.

For olefin (and acetylene) metathesis reaction in equation *3a*

(3a) $R^1HC = CHR^4 + R^2HC = CHR^3 \rightarrow R^2HC = CHR^1 + R^3HC = CHR^4$

Chauvin [2] proposed a mechanism involving the formation of a metallacycle, **3**, equation *3b*.

$$L_nM{=}CHR^1 + R^2CH{=}CHR^3 \longrightarrow$$

(3b) $\begin{array}{ccc} R^1HC & \text{—} & CHR^2 \\ | & & | \\ L_nM & \text{—} & CHR^3 \end{array} \longrightarrow L_nM{=}CHR^3 + R^2CH{=}CHR^1$

3

More recently Watson [3] and Parshall have discovered the σ-bond metathesis process of importance for the activation of C-H bonds:

(4) $L_nM\text{-}R^* + H\text{-}R \longrightarrow \begin{array}{ccc} R^* & \text{- - -} & H \\ \vdots & & \vdots \\ L_nM & \text{- - -} & R \end{array} \longrightarrow L_nM\text{-}R + H\text{-}R^*$

4

The reactions in equations *2–4* have already been studied extensively by theoretical model calculations due to Rappé [4], Morokuma [5, 6], Marynick [7], Hoffmann [8], Fujimoto [9], Ziegler [10], Goddard [11] and others.

We shall here make use of the most recent generation of density functionals [12] in a systematic study on four center transition states and intermediates in organometallic chemistry. The speed and accuracy of the new density functionals allow us to carry out realistic calculations on the structure and energies of even larger size systems. We shall in addition try to explain why the $[2_s + 2_s]$ organometallic addition reactions in equations 2–4 are feasible whereas the analogous $[2_s + 2_s]$ organic addition reactions [13] are subject to an insurmountable barrier. Our investigation will cover the sigma-bond metathesis reaction of equation *4* in Section 3, whereas the double bond metathesis reaction of equation *3b* will be dealt with in Section 4. We shall finally deal with triple bond metathesis in Section 5.

2. Computational details

The reported calculations were all carried out by utilizing the HFS-LCAO program system A-MOL, developed by Baerends *et al.* [14, 15] and vectorized by Ravenek. [16] The numerical integration procedure applied for the calculations was developed by teVelde *et al.* [17]. The geometry optimization procedure was based on the method developed by Versluis [18] and Ziegler. The electronic valence configurations of the molecular systems were described by the uncontracted triple-ζ STO basis sets [19] on metals for ns, np, nd, (n + 1)s and (n + 1)p shells as well as a double-ζ STO basis set [19] on carbon (2s, 2p), chlorine (3s, 3p) and hydrogen (1s). Hydrogens and carbons were given an extra polarization function: $3d_C$ (ζ_{3d} = 2.5); $2p_H$ (ζ_{2p} = 2.0). The frozen-core approximation was applied. A set of auxiliary [20] s, p, d, f and g STO functions, centered on all nuclei, was used in order to fit the molecular density and present Coulomb and exchange potentials accurately in each SCF cycle. Energy differences were calculated by including the local exchange-correlation potential by Vosko [21] *et al.* with Becke's [22] non-local exchange corrections and Perdew's [12b] nonlocal correlation correction. Geometries were optimized without including nonlocal corrections. The application of approximate density functional theory to organometallic chemistry has been reviewed recently [23e].

Approximate Density Functional Theory [24], which over the past decade has emerged as a tangible and versatile computational method, has been employed successfully to obtain thermochemical data [25, 26]; molecular structures [27, 28]; force fields and frequencies [29]; assignments of NMR- [30, 31], photoelectron- [32], ESR [33]-spectra, transition state structures as well as activation barriers [34]; dipole moments [35] and other one-electron properties. Thus, approximate DFT is now applied to many problems previously covered exclusively by *ab initio* Hartree-Fock (HF) and post-HF methods. The recently acquired popularity of approximate DFT stems in large measure from its computational expedience which makes it amenable even to large size molecules at a fraction of the time required for HF or post-HF calculations. More importantly, perhaps, is the fact that expectation values derived from approximate DFT in most cases are better in line with experiment than results obtained from HF calculations. This is in particular the case for systems involving transition metals. An analysis of why approximate DFT affords more reliable results than HF has recently been published by Cook [36] and Karplus as well as Tschinke and Ziegler [37]. All reported equilibrium structures were confirmed to represent energy minima by considering the force constant matrix over the optimization variables.

3. Single bond addition

3.1. INTRODUCTION

The ability of metal centers to mediate the breaking and formation of H-H and C-H bonds is of considerable practical importance. Thus, several experimental [3, 38–41] and theoretical [7, 42–45] studies have been undertaken in order to understand these pivotal elementary steps. The initial investigations focused on electron rich middle to late transition metal centers which can activate (break) H-H [38, 42] and C-H [39, 42i–k, 43a–c] bonds by oxidative addition and subsequently generate new H-H and C-H bonds by reductive elimination. Of crucial importance in these types of reactions is the ability of the metal center to change its formal oxidation state. More recent studies have been directed towards the ability of early electron poor metal centers to break and form H-H and C-H σ bonds through the σ-bond metathesis like reaction [3, 40]

(5) $L_nM\text{-}R^* + H\text{-}R \rightarrow L_nM\text{-}R + H\text{-}R^*$
($M = d^0, f^{14}d^0$; R^* = H, alkyl, R = H, alkyl,alkenyl,alkynyl)

where the formal oxidation state and electron count on the metal center remains unchanged. The σ-bond metathesis [40h] by organo-lutetium and organo-scandium methyl complexes have been studied experimentally by Watson [3] and Bercaw [40a–b] in connection with the Ziegler-Natta polymerization process. Electron poor early f-block elements are also known [40c–d] to activate H-H and C-H bonds according the reaction given in equation *5*.

Activation of the H-H and C-H bonds by electron poor metal centres in the σ-bond metathesis reaction of equation *4* have been postulated [3, 40] to proceed via a 4-centre transition state **4**. Activation of H-H bonds via the four centre transition state, **4**, might also occur in complexes of middle to late transition metals [45].

Up to date theoretical investigations on σ-bond metathesis include a GVB (Generalized Valence Bond) study of the reaction in equation *5* (M = Sc, Ti; L = Cl) with R^* = H; R = H by Goddard and Steigerwald [44a] and EHA (Extended Hückel Approximation) calculations of the reaction in equation *5* (M = Lu, L = Cp^*) with R^* = H; R = H and $R^* = CH_3$; $R = CH_3$ by Hoffmann and co-workers [8]. Recently, Rappé [4b] has discussed σ-bond metathesis between C-H bonds of acetylenes and the $Cl_2Sc\text{-}R$ (R = H, CH_3) linkage. We [44b–c] have also reported preliminary calculations on the reaction in equation *5* with $ML_n = LuCl_2$ and R^* = H, CH_3; R = H, CH_3.

The present investigation is concerned with $Cp_2Sc\text{-}R$ (R = H, alkyl, alkenyl, and alkynyl) as well as the ability of Cp_2Sc-H and Cp_2Sc-CH_3 to undergo σ-bond metathesis reactions with C-H bonds of methane, ethylene and acetylene.

3.2. SIGMA-BOND METATHESIS REACTIONS INVOLVING Cp_2Sc-H AND Cp_2Sc-CH_3

We shall in the following section discuss the general σ-bond metathesis reaction of equation *6* where A-B represents a Sc-H or Sc-C σ-bond whereas C-D involves H-H or C-H single bonds. This type of reactions has been studied for scandium and other electron poor metal centers in a number of experimental [40a–b, g] and theoretical [8, 44a] investigations following the pioneering work by P. Watson [3].

$$(6) \quad \begin{matrix} A \\ | \\ B \end{matrix} + \begin{matrix} C \\ | \\ D \end{matrix} \longrightarrow \begin{matrix} A — C \\ + \\ B — D \end{matrix}$$

We have optimized the structures of the two organometallic reactants: Cp_2Sc-H, **5a**, and Cp_2Sc-CH_3, **5b**, under C_s symmetry constraints. We did not find any energy minimum for the methyl complex with an agostic interaction.

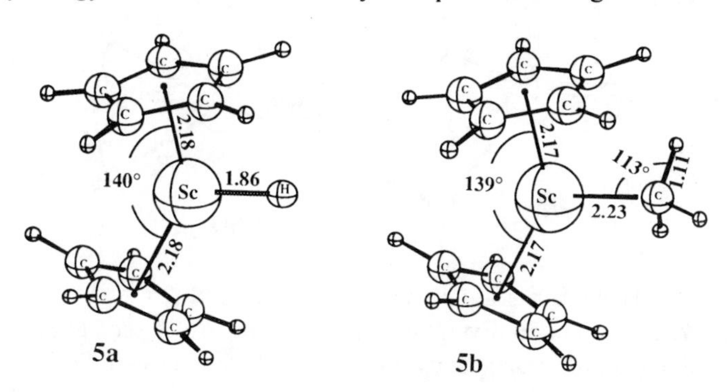

We have also optimized the geometries of the ethyl and propyl complexes and established, that the most stable conformations are respectively the eclipsed, single β-agostic structure **5c** and the $(C_\alpha\text{-}C_\beta)_{eclipsed}$, $(C_\beta\text{-}C_\gamma)_{staggered}$ structure **5d**.

3.2.1. *Activation of the H-H bond*

The hydrogen exchange reaction of equation *7* has been observed experimentally [40b, 41]. It is exceedingly fast compared to other types of σ-bond metathesis reactions involving scandium as well as other electron poor metals.

(7) $Cp_2Sc\text{-}H + D\text{-}D \rightarrow Cp_2Sc\text{-}D + H\text{-}D$

Our schematic reaction profile is given in Figure *1a*. We find that the degenerate metathesis process in equation *7* proceeds from reactants over a weakly bound dihydrogen adduct **6a** to the four center transition state **6b** with a kite-shaped Sc-H-H-H core. The adduct formation energy for **6a** amounts to 16 kJ mol^{-1} and the intermediate **6a** is characterized by a H-H bond which is stretched .04 Å compared to free H_2.

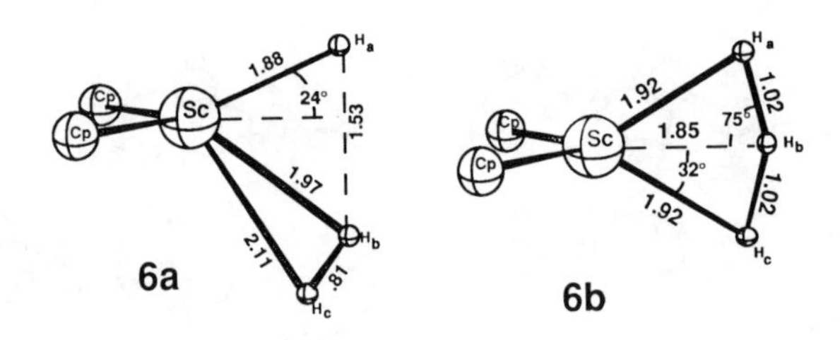

The two Sc-H distances are respectively .11 Å and .25 Å, longer than the regular Sc-H bond in **5a**. Even the transition state, **6b**, is seen to be modestly more stable than the reactants by 7 kJ mol^{-1}. The kite-shaped Sc-H-H-H core in **6b** exhibits three Sc-H contacts of nearly the same lengths as the Sc-H hydride distance in **5a** and two H-H bonds that have been stretched by .25 Å compared to free H_2. Thus a considerable amount of bond making and bond breaking is taking place in **6b**. On the whole, the loss in H-H interaction energy experienced by **6b** is compensated for by the formation of two new Sc-H links so that the transition state is of nearly the same energy as the reactants. It can be concluded from Figure *1a* that the hydrogen exchange reaction of equation *7* will proceed readily without any barrier. In fact, the entire potential surface connecting $Cp_2Sc\text{-}H + D_2$ with Sc-D and Sc-D + HD

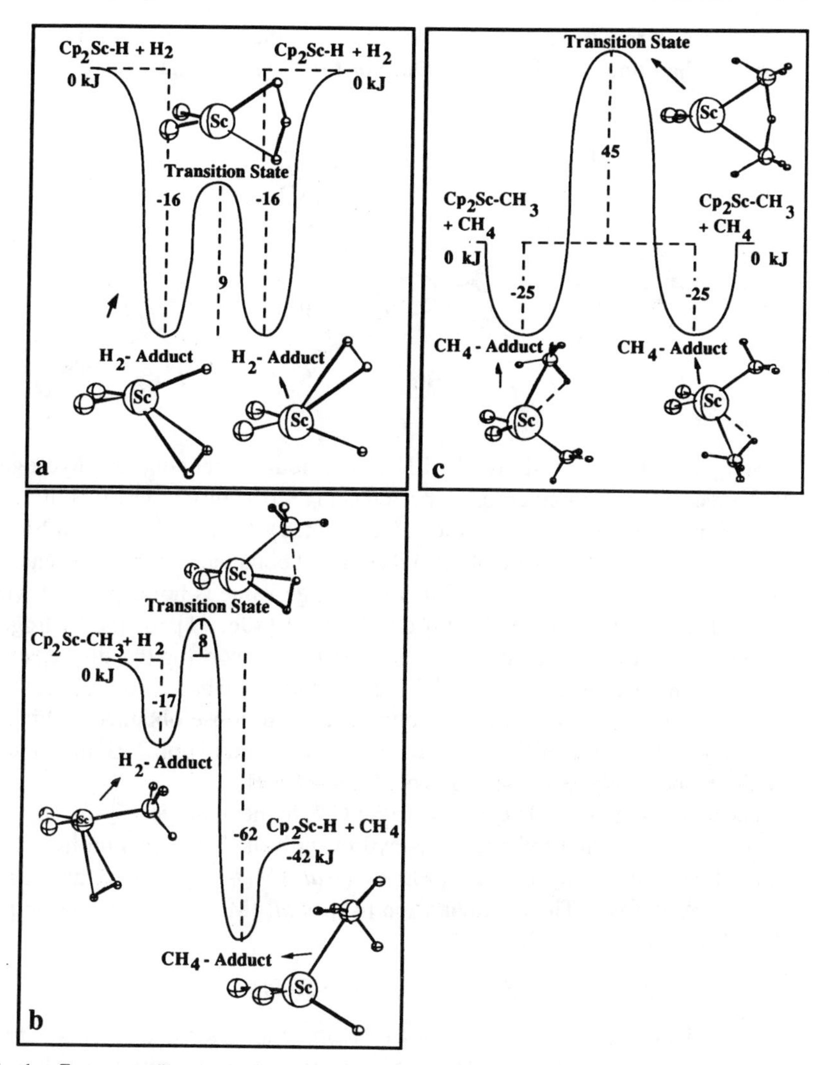

Fig. 1. Energy profiles for σ-bond metathesis reactions. All energies in kJ/mol. (a) equation *7*. (b) equation *8*. (c) equation *10*.

is so shallow that substantial geometrical distortions of **6a** and **6b** can be accomplished essentially without changing the total energy. It is thus likely that individual reaction paths can proceed through structures that deviate considerably from **6a** and **6b**. The force constant matrix over the optimization variables had one negative eigenvalue for the transition state **6b**.

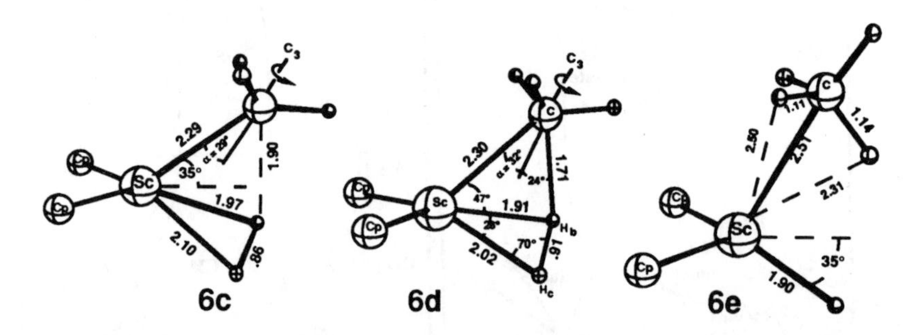

Steigerwald [44a] and Goddard have previously modelled the hydrogen exchange reaction of equation *7* by replacing Cp with Cl. Their optimized transition state has essentially the same geometry for the kite-shaped Sc-H-H-H core as **6b**. Steigerwald [44a] and Goddard obtained an activation energy of 71 kJ mol^{-1} (17 kcal mol^{-1}). This rather large value is not consistent with the high rates of 10^3 s^{-1} M^{-1} (–90°C) observed [40b, 41] for the hydrogen exchange reaction in equation *7*. One possible reason for the discrepancy might be the use of Cl instead of Cp. However, DFT-based calculations on the chloro-system afforded results quite similar to those obtained in Figure *1a*. Thus with Cl as a co-ligand **6a** and **6b** are 19 kJ mol^{-1} and 7 kJ mol^{-1} more stable, respectively, than the reactants Cl_2Sc-H + H_2.

The activation of a H-H bond by Cp_2Sc-CH_3 in the hydrogenolysis reaction of equation *4* is often referred to as hydrogenolysis. This reaction has been studied extensively by Bercaw [40a–b] *et al.* (M = Sc) as well as Marks [40c–d] *et al.* (M = Th,U), Richardson [41] *et al.* (M = Zr^+) and Watson [3] (M = Lu). It is very facile.

$$(8) \qquad Cp_2Sc\text{-}CH_3 + H\text{-}H \rightarrow Cp_2Sc\text{-}H + H\text{-}CH_3$$

Our calculated energy profile for the hydrogenolysis reaction of equation *8* is displayed in Figure *1b*. The profile passes from an initial dihydrogen adduct, **6c**, over the transition state, **6d**, to the product like adduct between methane and Cp_2Sc-H, **6e**. The reaction exhibits a modest barrier of 8 kJ mol^{-1} and an exothermicity of 42 kJ mol^{-1}.

The structure of the transition state **6d** is reached early in the process and resembles closely the H_2 adduct, **6c**, with a stretched H-H bond (.91Å) and two close Sc-H contacts of 1.97 Å and 2.11 Å, respectively, **6d**. The transition state is destabilized relative to the reactants by the H-H stretch as well as the tilt of the local C_3 axis on the methyl group away from the Sc-C bond vector towards the direction of the incoming hydrogen, **6d**. The tilt of the C_3 axis in **6d** by an angle α of 32° redirects the σ-orbital on CH_3 towards the incoming H-atom at the expense of losing some bonding interaction with the metal based orbitals on scandium, thus weakening the Sc-C bond. The Sc-H_a interaction in the transition state for the hydrogen exchange reaction, **6b**, does not suffer a similar destabilization since [40b, 42] the spherical nature of the $1s_a$ orbital is well suited for maintaining a strong Sc-H_a interaction while the H_a-H_b bond to the incoming H_b atom is formed. It is thus understandable that the hydrogenolysis reaction has a lower barrier than the hydrogen exchange reaction. Several authors [40b, 42] have pointed out that the 1s-hydrogen orbital is better suited to stabilize four center transition states, **6b**, than the directional methyl σ-orbital. The small calculated barrier of 8 kJ mol^{-1} for the hydrogenolysis reaction is in harmony with experimental observations [40b] according to which the reaction is fast with a rate constant of 4 × 10^{-1} s^{-1} M^{-1} at –78°C. The force constant matrix over the optimization variables of **6d** was confirmed to have one negative eigenvalue.

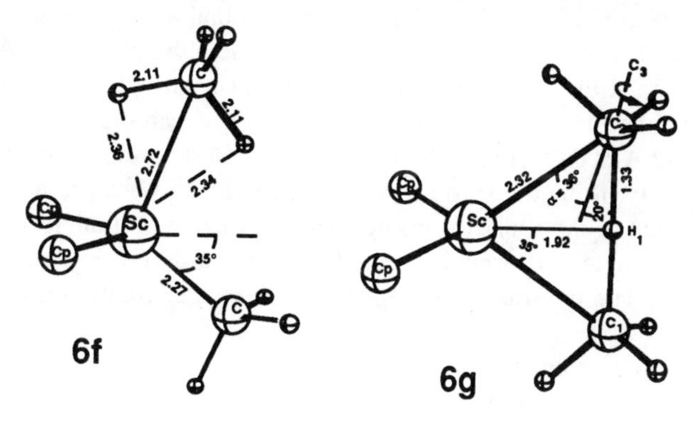

3.2.2. *Activation of the alkylic C-H bond*

The reverse of the hydrogenolysis reaction given in equation *8* represents an example where the alkylic C-H bond has been activated by Cp_2Sc-H in a σ-bond metathesis reaction, equation *9*.

(9) $Cp_2Sc\text{-}H + H\text{-}CH_3 \rightarrow Cp_2Sc\text{-}CH_3 + H\text{-}H$

It follows from our calculations that the methylation reaction in equation *9* is endothermic by 42 kJ mol^{-1}. Figure *1b*. This is in line with experimental studies where Thompson [40b] *et al.* as well as Christ [41] *et al.* find that hydrogenolysis, equation *8*, is more facile than methylation, equation *9*, with the equilibrium in equation *9* shifted to the left. The endothermicity of equation *9* is due to the fact that a Sc-H bond is stronger than a Sc-CH_3 bond by 45 kJ mol^{-1}, Table *II*.

The methane exchange process of equation *10*

$$(10) \qquad Cp_2Sc\text{-}C^*H_3 + H\text{-}CH_3 \rightarrow Cp_2Sc\text{-}CH_3 + H\text{-}C^*H_3$$

was discovered by Watson [3a] with lutetium as the metal atom. This process represents the first known example of methane activation by an organometallic complex. Alkyl exchange reactions have since been studied by Thompson [40b] *et al.* (M = Sc), Christ [41] *et al.* (M = Zr^+) and Marks [40c–d] *et al.* (M = Th,U).

The calculated energy profile for the methane exchange reaction is displayed in Figure *1c*. The process has a barrier of 45 kJ mol^{-1} and a four center transition state, **6g**, of C_{2v} symmetry. A methane adduct, **6f**, is formed in the early stages of the reaction with a formation energy of 25 kJ mol^{-1}. The transition state structure, **6g**, reveals that the activated C_1-H_1 bond has been stretched to 1.33 Å. The stretch is compensated for by the formation of new bonds involving C_1 as well as H_1. Thus, C_1 has formed a Sc-C_1 bond with a Sc-C distances of 2.31 Å, which is only slightly longer than the Sc-C bond of 2.26 Å in **5b**. Hydrogen has at the same time established a full bond with scandium and a weaker stretched H_1-C_2 bond to the adjacent methyl carbon. The Sc-H_1 distance of 1.92 Å is only slightly longer than the hydride distance in **5a**.

The stability of the transition state **6g** is hampered by the presence of two methyl groups in the four-center kite shaped core. The tilt angle α in **6g** is 36° with the methyl σ-orbital directed more towards H_1 than towards the metal center. The two methyl groups have a near free rotation around their respective C_3-axis.

The C-H activation in the methyl exchange reaction, equation *10*, is observed experimentally to proceed at a much slower rate than the H-H activation processes of equations *7* and *8*. Watson[46] finds an activation energy of 49 kJ mol^{-1} for the methyl exchange reaction involving lutetium which is quite close to our estimated value for the electronic barrier in the scandium system at 45 kJ mol^{-1}, Figure *1c*. The force constant matrix over the optimization variables of **6g** was confirmed to have one negative eigenvalue.

We have shown that the electronic energy barrier for a σ-bond metathesis reactions increases as we increase the number of methyl groups in the kite shaped core of the four-center transition state. We shall now discuss how vinyl and acetylide groups will influence the stability of a four center transition state in connection with a study of σ-bond metathesis involving C-H alkenyl and alkynyl bonds.

3.2.3. *Activation of the alkenylic C-H bond*

Alkenes can react with L_2M-R (R = H, CH_3, L = Cp*, M = Lu, Sc, Zr^+ and Th) by insertion into the M-R bond

(11) $L_2M\text{-}R + H_2C = CR'R'' \rightarrow L_2M\text{-}CH_2\text{-}CRR'R''$
(a) R = H.
(b) R = CH_3.

or alternatively by an activation of the alkenylic C-H bond in a σ-bond metathesis reaction

(12) $L_2M\text{-}R + H_2C = CR'H \rightarrow L_2M\text{-}C(H) = CR'R'' + H\text{-}R$

For lutetium the simplest and least sterically hindered olefins, ethylene and propene, react by insertion [46] whereas more bulky olefins give rise to C-H activation. For scandium [40a] only ethylene inserts into the Sc-R bond whereas more bulky olefins than propene can insert into M-R bonds of actinides. The trend has been rationalized [46, 40a] by observing that the transition state for the insertion becomes sterically crowded with more bulky olefins, in particular for the smaller sized metals such as scandium. The four center transition state for C-H activation is sterically less demanding and thus accessible to bulkier olefins. It has further been observed that C-H activation in olefins invariably takes place at the stronger alkenylic C-H bond rather than the weaker alkylic C-H bond. We shall first discuss C-H activation, equation *13*.

(13) $Cp_2Sc\text{-}H + H_2C = CH_2 \rightarrow Cp_2Sc\text{-}C(H) = CH_2 + H\text{-}H$

The reaction profile for activation of a vinylic C-H bond in ethylene by Cp_2Sc-H is given in Figure *2a*. The process is endothermic by 16 kJ mol^{-1} and has a modest barrier of 20 kJ mol^{-1}. The incoming ethylene forms an adduct, **7a**, which proceeds to the four-center transition state, **7b**. There is a clear resemblance between the transition states corresponding to the activation by Cp_2Sc-H of an alkylic C-H bond, **6d**, and a vinylic C-H bond, **7b**. Both structures have a largely broken C-H bond. Further, the C-H bond breaking is compensated for by the formation of Sc-C and H-H bonds.

The correlation diagram for the two types of C-H activation by Cp_2Sc-H is illustrated in schematic form in Figure *3a*. There are two electron pairs in-

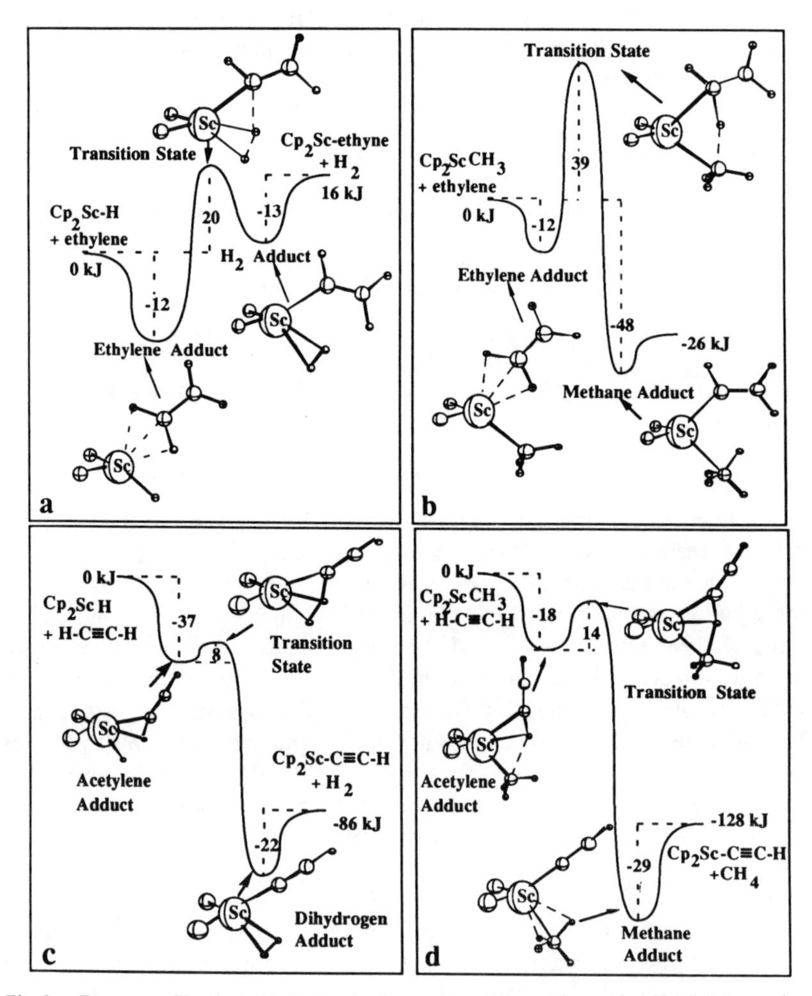

Fig. 2. Energy profiles for σ-bond metathesis reactions. All energies are in kJ/mol. (a) equation 13. (b) equation *14*. (c) equation *17a*. (d) equation *19*.

volved in the process. The pair of lowest energy is originally situated in a C-H σ-orbital, right hand side of Figure *3a*, which correlates smoothly with the H-H σ-orbital on the product side, left hand side of Figure *3a*. The electron pair is in the transition state extended over the α-carbon center as well as the two hydrogen atoms, with the largest amplitude on the hydrogens. The orbitals

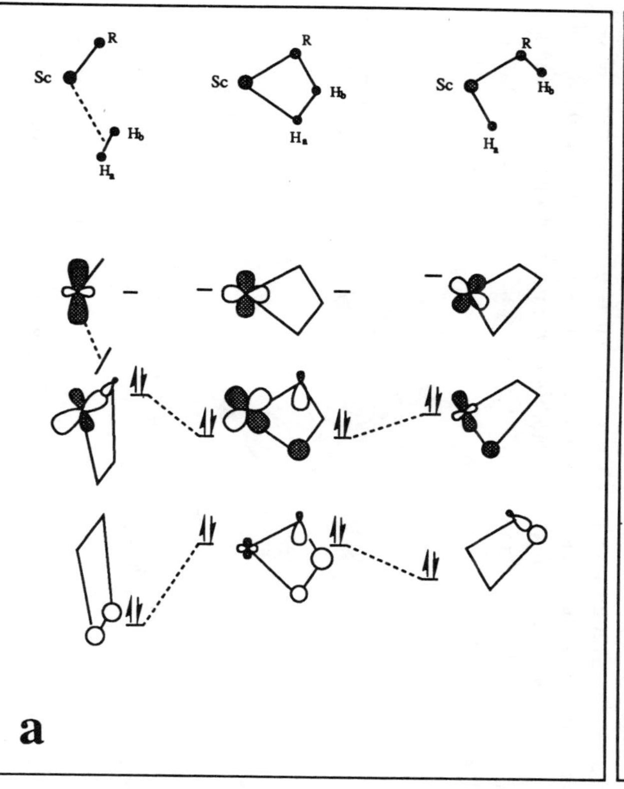

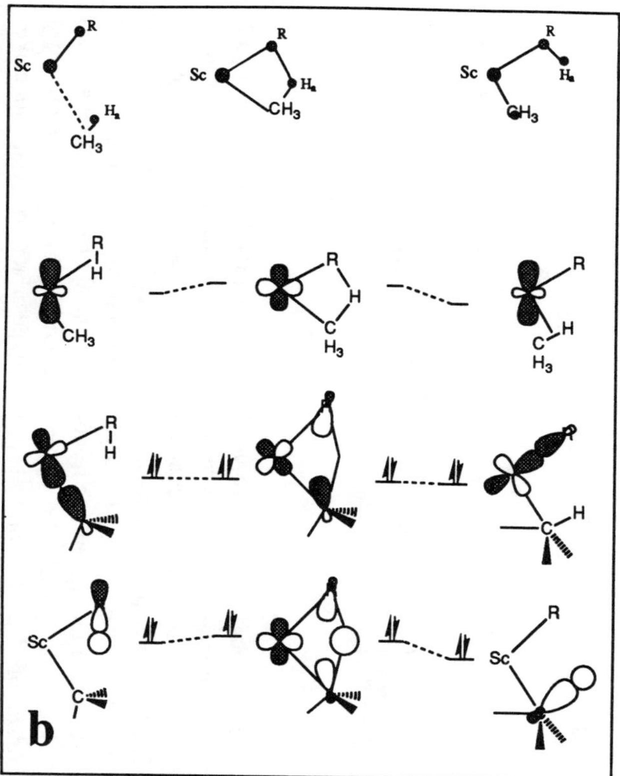

Fig. 3. Schematic correlation diagram for σ-bond metathesis reactions. (a) equation *7*. (b) equation *10* and *12*; R = alkyl, vinyl.

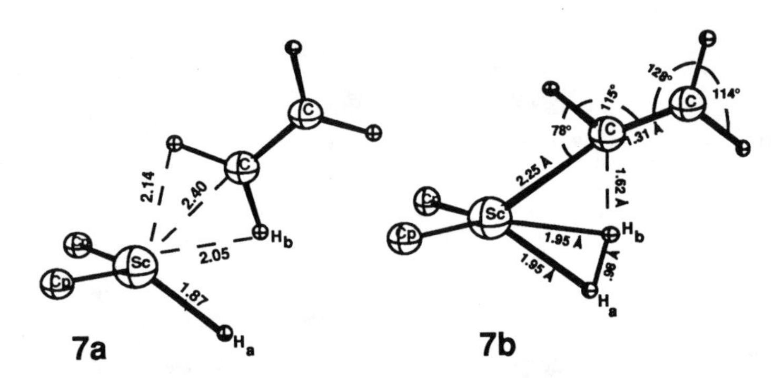

on the metal center are not involved in stabilizing the lower electron pair. The pair of highest energy resides on the reactant side in a Sc-H σ-bond which correlates on the product side with a Sc-C σ-orbital. The transition state has the upper pair situated in a three center orbital involving d_{xy} on scandium as well as the σ_C on the α-carbon and $1s_H$ on H_a. The upper electron pair is stabilized throughout the reaction by a metal orbital which maintains optimal overlaps with the adjacent ligand orbitals by a constant rehybridization. A pool of empty s, p, and d-type orbitals on scandium makes such a rehybridization possible. The d_{xy} orbital is of particular importance for the stability of the electron pair in the transition state, whereas d_{xy} as well as $d_{x^2-y^2}$ play a pivotal role for the stability of the upper electron pair among the products and the reactants. The C-H bond breaking transit from **7a** to **7b** is illustrated in **7c**.

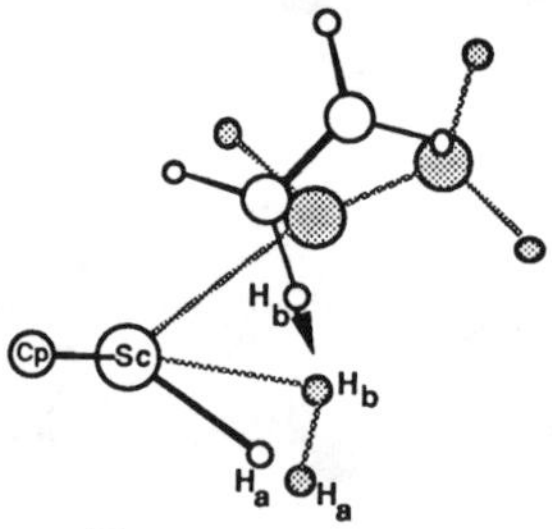

Cp Sc H

7c

The energy profile for the activation of a vinylic C-H bond by $Cp_2Sc\text{-}CH_3$ in equation *14* is given in Figure *2b*.

$$(14) \qquad Cp_2Sc\text{-}CH_3 + H_2C = CH_2 \rightarrow Cp_2Sc\text{-}C(H) = CH_2 + CH_4$$

The reaction is exothermic by 26 kJ mol^{-1} and subject to an electronic activation barrier of 39 kJ mol^{-1}. Thus vinylic C-H activation (equation *14*; Figure *2b*) mediated by $Cp_2Sc\text{-}CH_3$ is seen to be exothermic unlike the corresponding process mediated by Cp_2Sc-H (equation *13*; Figure *2a*). However, vinylic C-H activation by $Cp_2Sc\text{-}CH_3$ faces a larger barrier than vinylic C-H activation involving Cp_2Sc-H.

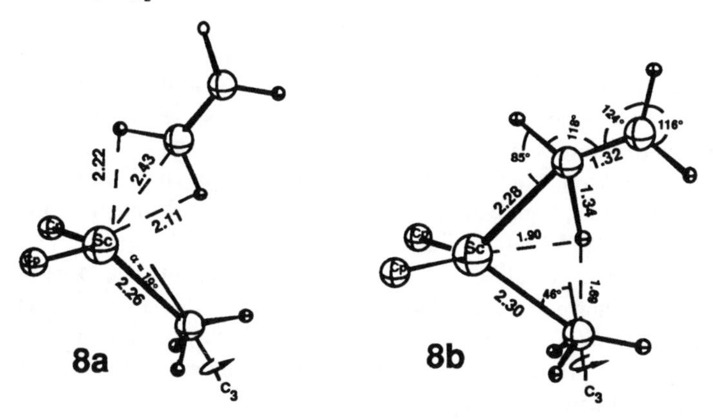

The approximate transition state, **8b**, exhibits a vinylic C-H bond that has been stretched to 1.34 Å as well as a methyl group for which the tilt angle α has been increased from 19° in the ethane adduct, **8a**, to 46° in the transition state. The two destabilizing factors are encountered by the formation of a Sc-H bond at 1.90 Å, **8b**, as well as a vinylic Sc-C bond at 2.30 Å. The transition state for the alkylic C-H activation by $Cp_2Sc\text{-}CH_3$, **6g**, is in many ways similar to that, **8b**, corresponding to the vinylic C-H activation by $Cp_2Sc\text{-}CH_3$. Both **6g** and **8b** are destabilized by the fact that the two carbon centers in the four-center core have directional σ-orbitals which are unable to sustain optimal interactions with hydrogen and the metal center at the same time. The result is a substantial electronic barrier of ~40 kJ mol^{-1}. The correlation diagram for the two types of C-H activation by $Cp_2Sc\text{-}CH_3$ is illustrated in schematic form in Figure *3b*. The correlation diagram is very similar to that presented in Figure *3a* for the corresponding C-H activation processes mediated by Cp_2Sc-H. There is again two electron pairs involved in the process and both are stabilized by the ability of the metal center to rehybridize throughout the reaction in order to maintain overlaps with the two adjacent σ-orbitals on the carbon centers. Again, the directionality of the σ-orbitals impedes optimal simultaneous overlaps with the 1s-orbital and the metal hybrides in the two orbitals holding the active electron pairs. The energy barrier for alkenylic C-

H activation by Cp*Sc-CH_3 has been measured [40b] for p-methoxystyrene as 48 kJ mol^{-1}, which is not substantially different from our calculated electronic barrier of 39 kJ mol^{-1} for C-H activation in ethylene (Figure *2a*). The system in **8b** was too large for a calculation of the force constant matrix and we were thus not able to verify whether **8b** has a single negative eigenvalue.

Thompson [40b] *et al.* have observed that olefins containing vinylic as well as alkylic C-H bonds undergo σ-bond metathesis with Cp*Sc-CH_3 preferably using the vinylic C-H bond. This preference could be ascribed [40b] to kinetic factors assuming that the less directional alkenylic sp^2 σ-orbital is better able to stabilize the four center transition state than the sp^3-type alkylic σ-orbital. However, we calculate the barrier for alkenylic C-H bond activation at 39 kJ mol^{-1}, Figure *2b*, to be only slightly lower than that of alkylic C-H activation at 48 kJ mol^{-1}, Figure *1c*.

The preference for alkenylic C-H bond activation might alternatively be attributed to thermodynamic factors. In fact, we calculate the reaction in equation *14* to be exothermic by –26 kJ mol^{-1} which would indicate that Sc-vinyl bonds are prefered over Sc-alkyl bonds. The enthalpy for the reaction in equation *14* can be written as

$$(15) \qquad \Delta H_{14} = [D(H\text{-}C_{vinyl}) - D(H\text{-}C_{methyl})] + [D(Sc\text{-}C_{methyl}) - D(Sc\text{-}C_{vinyl})]$$

The difference in the first square bracket is positive as vinylic C-H bonds are stronger than alkylic C-H bonds, Table *I*. Thus the first square bracket, which we calculate to be 17 kJ mol^{-1}, would favor alkylic C-H activation and shift the equilibrium to the left in equation *14*. The second difference amounts to –43 kJ mol^{-1} according to our calculations, Table *II*. It indicates that the vinylic Sc-C bond is substantially stronger than an alkylic Sc-C bond and favors vinylic C-H activation by shifting the equilibrium to the right in equation *14*. Our analysis leads to the conclusion that the preference for vinylic C-H bond activation is driven by the strength of the vinylic Sc-C bond.

We have previously calculated the reaction enthalpy for the insertion of ethylene into the Cp_2Sc-H bond as ΔH_{11a} = –115 kJ mol^{-1}, whereas the σ-bond metathesis reaction involving Cp_2Sc-H and ethylene, equation *13*, has a reaction enthalpy of ΔH_{13} = 16 kJ mol^{-1}, Figure *2a*. Thus, insertion is favored over σ-bond metathesis on thermochemical grounds for the reaction between Cp_2Sc-H and ethylene. Insertion is also calculated to be favored in the reaction between Cp_2Sc-CH_3 and the ethylene molecule. Here, the reaction enthalpy for insertion is given as ΔH_{11b} = –65 kJ mol^{-1}, whereas σ-bond metathesis gives ΔH_{14} = –26 kJ mol^{-1}, Figure *2b*. We shall in a later study [47] look at the reaction profiles for insertion processes and evaluate the influence that

TABLE I

H-R bond energies[i]

Molecule	Bond	Bond dissociation energies			
		Exp.		Calc.	
		D_e	D_0	D_e	D_0
H_2	H-H	463[a]	432[a]	472	441
CH_4	C-H	468[c]	431[b]	469	435
C_2H_6	C-H		416[d]	451	414
C_3H_8	C-H[e]			440	
C_2H_5	C_b-H[f]			179	
C_2H_4	C-H		459[b]	486	451
C_2H_2	C-H		549[b]	586	555
$CH_3CH_2CH_2$	C_g-C_b[g]			120	
$CHCH_2$	C_b-H[h]			193	

[a] Herzberg, G. In "Spectra of Diatomic Molecules" Van Norstrand Reinhold, New York, 1950.
[b] K.M. Ervin *et al.* J. Am. Chem. Soc., **112**, 5750 (1990).
[c] Ref. 46.
[d] B. Rusicic *et al.*, J. Chem. Phys., **91**, 114 (1989).
[e] C-H bond energy in propane for terminal carbon.
[f] C-H bond energy in ethyl radical for β-carbon.
[g] CH_3-CH_2CH_2 bond energy in propyl radical.
[h] C-H bond in vinyl radical for β-carbon.
[i] All energies in kJ mol^{-1}.

TABLE II

Sc-R bond energies[c]

Molecule	Bond	Bond dissociation energy
		Calc.
		D_e
Cp_2Sc-H	Sc-H	340
Cp_2Sc-CH_3	Sc-CH_3	295
Cp_2Sc-C_2H_5[a]	Sc-C_2H_5	283
Cp_2Sc-C_3H_7[b]	Sc-C_3H_7	240
Cp_2Sc-C_2H_3	Sc-C_2H_3	338
CpSc-C_2H	Sc-C_2H	540

[a] The single β-agostic, eclipsed conformation.
[b] The double γ-agostic, (C_α-C_β) eclipsed, (C_β-C_γ) staggered conformation.
[c] All energies in kJ mol^{-1}.

substituents [40a, 46] on either ethylene or cyclopentadienyl rings might have on the activation barrier.

3.2.4. *Activation of the alkynylic C-H bond*

Metal-hydride and metal-alkyl bonds of early transition metals can react [46] with alkynes via alkyne insertion, equation *16a*, or C-H activation, equation *16b*. Cp^*_2 Sc-CH_3 reacts exclusively [40b] with internal acetylenes to afford insertion products, equation *16a*, whereas C-H activation is observed exclusively in the reaction between Cp^*_2 Sc-CH_3 and terminal [40a, 46] acetylenes, equation *16b*. The preference observed for reactions between Cp^*_2Sc-CH_3 and alkynes indicates that this system favors alkynic C-H activation over insertion, whereas alkylic C-H activation is less favorable than either insertion, equation *16a* or alkynic C-H activation, equation *16b*.

(16a) $L_2M\text{-}R + R'C \equiv C\text{-}R'' \rightarrow L_2M\text{—}C(R'')\text{=}C(R)\text{—}R'$

(16b) $L_2M\text{-}R + R'C \equiv C\text{-}R'' \xrightarrow{R' = H} L_2M\text{—}C \equiv C\text{—}R'' + H\text{—}R$

The activation of the C-H bond in acetylene by Cp_2Sc-H

(17a) $Cp_2Sc\text{-}H + HCCH \rightarrow Cp_2Sc\text{-}CCH + H_2$

is calculated to be strongly exothermic with a reaction enthalpy of –86 kJ mol^{-1}, Figure *2c*. The reaction enthalpy can be written as

(17a) $\Delta H_{17a} = [D_e(Sc\text{-}H) - D_e(Sc\text{-}CCH)] - [D_e(H\text{-}H) - D_e(H\text{-}CCH)]$

and it follows from the data in Tables *I* and *II* that the reaction in equation *17a* is exothermic because the increase in strength in going from a Sc-H bond to a Sc-CCH bond is larger than the increase in going from a H-H bond to a H-CCH bond.

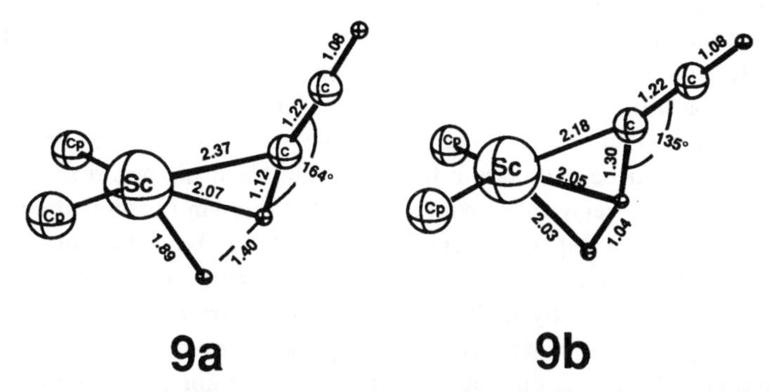

9a **9b**

Acetylene forms in the early stages of the reaction a tight adduct with Cp_2Sc-H, **9a**, in which a C-H bond on acetylene undergoes an agostic interaction with scandium. The agostic interaction results in an elongation of the C-H bond by .04 Å and a short Sc-H distance of 2.07 Å, **9a**.

The transition state, **9b**, for the process in equation *17a* has the C-H bond stretched to 1.30 Å. However, this bond weakening is more than compensated for by a new H-H bond with R(H-H) = 1.04 Å and a Sc-C bond of 2.18 Å which is comparable to the Sc-C bond length of the acetylide complex, **5b**. There is in addition a close Sc-H contact of 2.06 Å. The transition state is seen to be of lower energy than the sum of the two reactants Cp_2Sc-H and HCCH, Figure *2c*, although slightly above (8 kJ mol^{-1}) the adduct **9a**. It is clear from Figure *2c* that the activation of an alkynylic C-H bond should be a facile reaction from a thermodynamic as well as a kinetic points of view.

It might be of interest to compare the enthalpy ΔH_{17a} for the C-H activation with the enthalpy ΔH_{18a} for the alternative insertion process

(18a) $\quad Cp_2Sc\text{-}H + HCCH \rightarrow Cp_2Sc\text{-}C(H)CH_2$

The enthalpy ΔH_{18a} can be written as

(18b) $\quad \Delta H_{18a} = [D_e(Sc\text{-}H) - D_e(Sc\text{-}C(H)CH_2)] - D_e(H\text{-}C(H)CH)$

The difference from the first two terms in equation *18b* amounts to 2 kJ mol^{-1} as the Sc-H and Sc-C(H)CH_2 bonds are of nearly the same strength, Table *II*, whereas $-D_e$(H-C(H)CH) = −193 kJ mol^1, Table *I*. Thus insertion has an enthalpy of $\Delta H_{18a} = -191$ kJ mol^{-1}. The favorable reaction enthalpy stems primarily from the net formation of an additional σ-bond. It follows from our estimates in equations *17b* and *18b* that insertion should be favored over C-H activations on thermodynamic grounds in reactions involving scandium and we expect the same to be true for other early transition metals. Zirconium and hafnium [40f] do in fact favor insertion whereas both reactions are feasible with lutetium [3a]. For scandium [40b] only C-H activation is observed.

Rappé has discussed the reactions in equations *17a* and *18a* for a model system [4] in which the Cp-rings were replaced by Cl-atoms. Rappé found C-H activation to be exothermic by 63 kJ mol^{-1} and his optimized transition state has a four-center core with a geometry very similar to that of **9b**. The transition state in Rappé's calculation was found to be 25 kJ mol^{-1} above the reactants. Thus, both sets of theoretical studies point to the activation of an alkynylic C-H bond by hydrides of early transition metals as facile. However, Rappé finds in agreement with the present investigation, and counter to experimental observation [40b], that insertion, equation *18a*, is more favorable than C-H activation with the reaction enthalpy for insertion calculated at –153 kJ mol^{-1}. It has been suggested by Rappé and others [50] that the absence of insertion must be attibuted to kinetic factors. We intend to explore this point in a later study [47].

The profile for the corresponding activation of an alkynylic C-H bond by $Cp_2Sc\text{-}CH_3$ is displayed in Figure *2d*.

(19) $Cp_2Sc\text{-}CH_3 + HCCH \rightarrow Cp_2Sc\text{-}CCH + CH_4$

At the start of the reaction, we encounter again an acetylene adduct, **10a**, where the C-H bond which is going to be activated, is engaged in an agostic interaction with the electropositive metal center and stretched to 1.13 Å. The adduct stabilization energy is –18 kJ mol^{-1}.

The activated acetylic C-H bond is still retained in the transition state, **10b**, although, considerably weakened and stretched to 1.19 Å. The destabilization of the acetylic C-H bond is compensated for by the formation of Sc-H and Sc-C bonds at 1.97 Å and 2.23 Å, respectively. The stabilizing interaction between the electropositive scandium and the two centers on the acetylic C-H bond is primarily due to a donation of charge from the C-H bond to empty orbitals on the metal. The transition state is 4 kJ mol^{-1} below the reactants in energy, Figure *2d*, but still 14 kJ mol^{-1} above the acetylene adduct 10a. We were not able to verify whether **10b** has a single force constant matrix with a single negative eigenvalue due to the size of the system.

The C-H activation process in equation *14* is strongly exothermic with a calculated reaction enthalpy of –128 kJ/mol, Figure *2d*. It is thus clear that the process is feasible from a kinetic as well as thermodynamic point of view. In equation *19* a C-H methyl bond is formed at the expense of breaking a C-H acetylene bond which is energetically unfavorable. Table *I*. However,

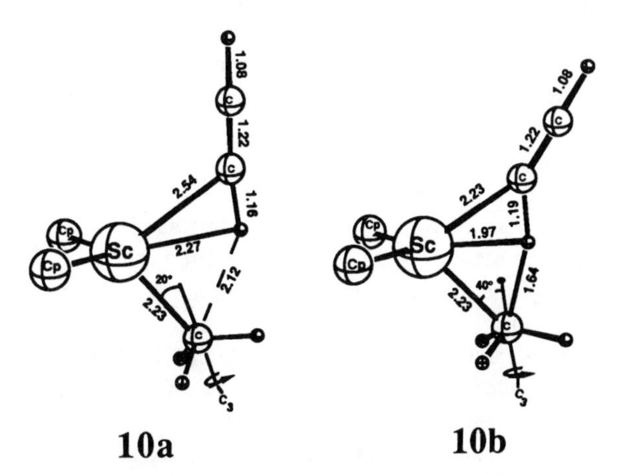

this loss in stability is more than compensated for by exchanging a Sc-methyl bond by a Sc-acetylide bond, Table *II*.

3.3. CONCLUDING REMARKS

We have studied various hydrocarbyl derivatives of scandocene. Most of the systems were found to possess a ground state structure in which the hydrocarbyl group is bound exclusively to scandium through a single carbon atom. The only exception was the ethyl derivative in which the Sc-C bond is supplemented by an agostic interaction between the metal and a β-hydrogen. The strength of the Sc-R bonds follow the expected trend alkynyl >> alkenyl > alkyl whereas the trend in bond strength among the alkyl species is methyl > ethyl > propyl. The calculated order is roughly the same as for the C-H bond in the corresponding H-R systems. We expect the calculated electronic bond dissociation energies to be accurate to within 30 kJ mol^{-1} and likely on the high side. Zero point energy corrections are not considered here for the Sc-R bonds. They should reduce the bond energies by 10 kJ mol^{-1}. The calculated order for the bond dissociation energies conform to the few trends observed experimentally [40g].

The second part of the study was concerned with the σ-bond metathesis reaction of equation *4*. The reaction was found to involve a four center transition state. The highest activation energies ~40 kJ mol^{-1} were obtained in the cases where two of the four groups in the core of the kite shape transition state structure, **4**, are alkyl or alkenyl, whereas the presence of a single alkenyl or alkyl group gives rise to a somewhat lower activation energy of 10 kJ

mol^{-1}. Processes involving only hydrides and alkynyl were found to have negative activation energies. The derived trends in activation energies follow closely the order in rates obtained experimentally [40, 46] and the increase in activation energy with alkyl or alkenyl groups can be understood, as suggested previously [5, 40a], from the directional nature of the σ-orbital on these groups which makes it impossible to maintain optimal interactions with both neighbours in the Sc-R-H-R′ core. It is shown that the formally forbidden $[_{\sigma}2_s + _{\sigma}2_s]$ reaction of equation *4* is made feasible by a pool of empty s, p, and d type orbitals on scandium which can supply suitable hybrides appropriate for optimal interaction with the neighboring groups in the Sc-R-H-R′ core throughout the reaction.

Our calculations indicate that insertion of ethylene, equations *11a* and *11b*, or acetylene, equation *18a*, into the Sc-H and Sc-CH_3 bonds are preferred thermodynamically over the alternative alkenylic, equations *13* and *14* or alkynylic, equations *17a* and *19*, C-H bond activations. This is in line with experiment for ethylene [40b]. However, acetylenes have been observed [40b] to prefer alkynylic C-H activation over insertion for scandium. We expect to investigate this point further by a full study [47] in which profiles for the insertion processes of olefins and acetylenes into the L_2Sc-R bonds (R = H, alkyl) are traced for L = Cp as well as methylated derivatives of Cp.

4. Double bond addition

4.1. INTRODUCTION

Carbene complexes of tungsten and molybdenum are efficient catalysts for the metathesis of olefin, equation *3a*. Chauvin [2] was the first to suggest that metal carbenes act as catalysts by undergoing a $[_{\pi}2_s + _{\pi}2_s]$ cycloaddition reaction with olefin, thus forming the metallacyclobutane, **11a**. The subsequent decomposition of an intermediate metallacycle **11a** results in the formation of a new olefin, equation *20*. Chauvin's mechanism is now widely accepted and several metallacyclobutanes have been isolated and characterized.

Schrock [48] *et al.* have studied the catalytic activity of $(OR)_2W(NR')CH_2$ and $(OR)_2Mo(NR')CH_2$. They find that alkoxy ligands with bulky and electron withdrawing substituents enhance the activity of $(OR)_2M(NR')CH_2$ whereas less electron withdrawing R groups such as R = t-Bu afford catalysts with little or no activity. Schrock [48] *et al.* have also carried out detailed investigations on the intermediate metallacycle **11a**. The active metal carbenes with bulky and electron withdrawing alkoxy groups such as $OCMe(CF_3)_2$

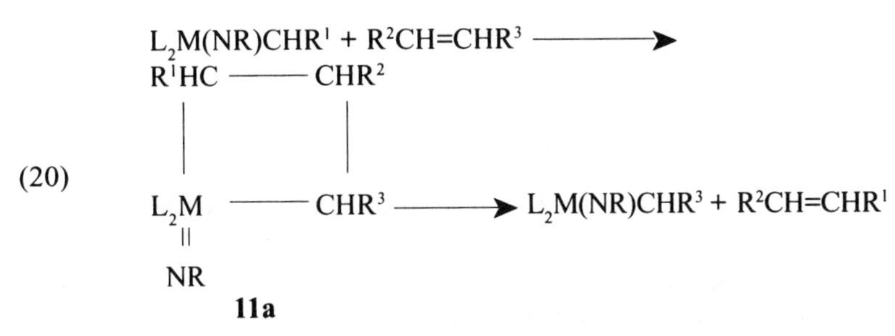

and $OC(CF_3)_2(CF_2CF_2CF_3)$ afford metallacycles with a trigonal bipyramidal structure, **11b**, whereas the less active carbenes with electron releasing alkoxy groups give rise to square pyramidal metallacycles, **11c**. The relation between the electronic properties of the alkoxy group, OR, and the activity of the carbene $(OR)_2M(NR')CH_2$ as well as the conformational preference of the corresponding metallacycle **11a** is not fully understood [48c].

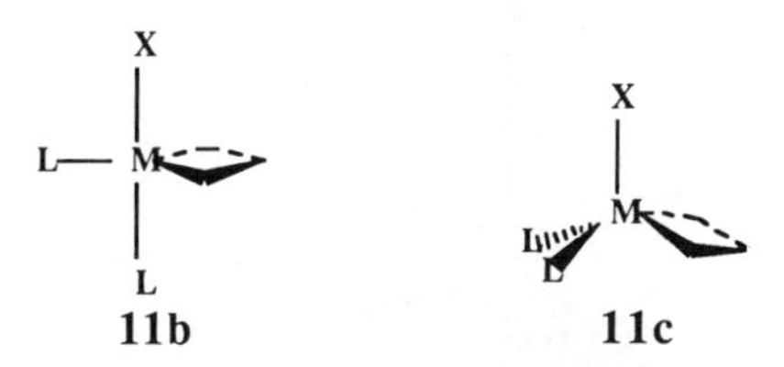

We present here a systematic study on the $[_{\pi}2_s + _{\pi}2_s]$ cycloaddition reaction of equation *20*, the key step in olefin metathesis catalyzed by $(L)_2Mo(X)CH_2$. The first objective of our investigation has been to understand why the formally symmetry forbidden $[_{\pi}2_s + _{\pi}2_s]$ cycloaddition reaction in equation *20* is facile whereas the corresponding addition reaction between two olefins, equation *3a*, has a very high activation barrier. We shall to this end study the electronic and molecular structures of the catalyst $(L)_2Mo(X)CH_2$ as well as the intermediate **11a**.

The up-to-date detailed calculations on related systems have been performed essentially by two groups, both of which have focused on early transition metal complexes. Using the EHT method, Hoffmann and co-workers [49] studied $Cp_2TiC_3H_6$, while Rappé and Upton [50a-b] utilized the GVB method to study the $Cl_2TiC_3H_6$ model system. Using the latter method, Rappé [50c-d] and Goddard have also calculated the reactivities of certain Cr, W and Mo dichloride methylidene systems towards olefins in studies on olefin metathesis and olefin polymerization. Cundari and Gordon [51] have more recently provided a theoretical analysis of metal carbenes used in olefin metathesis.

4.2. FORMATION AND DECOMPOSITION OF METALLACYCLES

We have carried out a detailed analysis of the process in equation *20* where the metal carbene of the type **12** reacts with ethylene to form the metallacycle **11c** with a square-pyramidal (SP) conformation as well as **11b** in a trigonal-bipyramidal (TBP) arrangement.

In our analysis below, we have used Cl to model real life alkoxy substituents. For one reason, chlorine lies in between methoxy and trifluoromethoxy ligands on the electrodonicity scale [52]. For another, it was computationally more economical to carry out such detailed calculations using ligands with a smaller electron count. We shall refer to the chloro- substituted metallacycles of SP and TBP geometries as **14** and **13**, respectively.

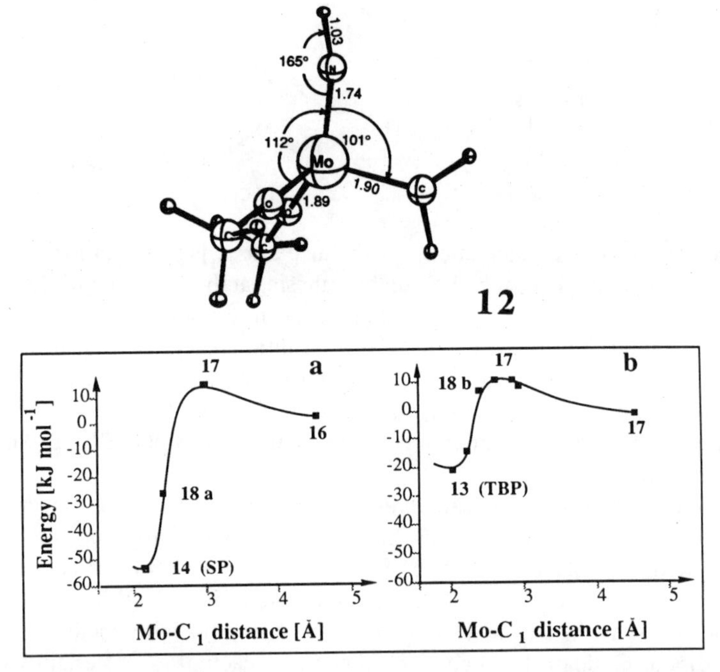

Fig. 4. Reaction profiles for the dissociation of the $Cl_2Mo(O)C_3H_6$ metallacycles (a) square-pyramidal **14**; (b) trigonal-bipyramidal **13**.

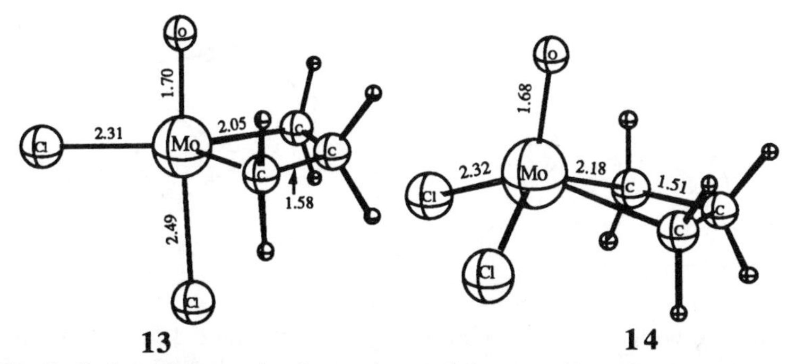

We shall first discuss the formations of the metallacycle **14** with a SP conformation. For this process, ethylene is allowed to approach the metal carbene by gradually decreasing the R(Mo-C_1) distance, **15a**, between the molybdenum center and one of the olefin carbons. The reaction coordinate R(Mo-C_1) was varied from 4.5 Å, where ethylene and the metal carbene virtually are non-interacting to 2.18 Å, which is the R(Mo-C_1) distance in the final metallacycle **14** of SP geometry. The variation of R(Mo-C_1) was carried out in 6 steps and for each of these steps all other degrees of freedom were optimized. The energy of the reacting system is plotted in Figure *4a* as a function of the reaction coordinate R(Mo-C_1).

It follows from our analysis that olefin approaches **12** perpendicular to the O-Mo-C_3 (xy) plane of the metal carbene, **15a**, in such a way that the C_1-C_2 olefin and Mo-C_3 carbene bonds are co-planar, xz of **15a**. The position of the C_1-C_2 olefin and Mo-C_3 carbene bonds are shown in **15b** for different values of the reaction coordinate R(Mo-C_1) in **16**, **17**, **18a** and **14**.

The position of Cl_1, Cl_2 and the spectator oxygen hardly changes during the reaction. The minor adjustments of the $Cl_2Mo(O)$ framework are not indicated in **15b**. The energy for each of the *en route* structures **16**, **17** and **18a** are given in Figure *4a* where we present a reaction energy profile for the process. There is virtually no interaction between ethylene and the metal carbene at the early stages, **16**, of the reaction where R(Mo-C_1) > 4 Å, and both reacting fragments are undistorted from the free state.

The energy is rising steadily from **16**, R(Mo-C_1) = 4.5 Å, to **17**, R(Mo-C_1) = 2.89 Å, by a total of 11.2 kJ mol^{-1}. The structure **17** can be considered as a transition state for the process in equation *20* since it is situated at the highest energy point on the approximate reaction profile, Figure *4a*. However, no attempt was made to verify whether **17** has a single normal mode with an imaginary frequency. The transition state structure **17** is basically an olefin

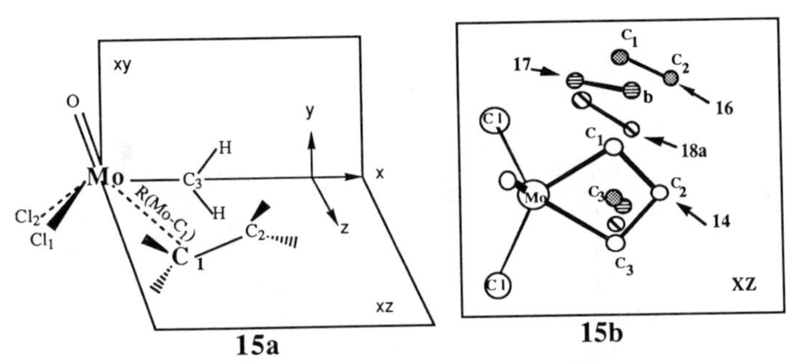

15a 15b

π-complex with an approximately trigonal-bipyramidal geometry. In this trigonal structure one of the chlorine ligands as well as the median of the olefin C-C bond are taking up the axial positions whereas the spectator oxygen, another chlorine ligand and the carbene carbon atom are lying in the equatorial plane of a bipyramid. The C=C bond distance in the weakly coordinated olefine was calculated to be 1.35 Å (versus 1.33 Å in a free molecule). At the same time, the hydrogen atoms in the olefin have started to bend away from the target carbene complex and the Mo-C carbene bond distances has increased from 1.88 Å to 1.90 Å as the carbene group moves out of the xy plane, **15a**.

The formation of the π-complex is followed by a steep decrease in the total energy until the product **14** is formed. All three carbon atoms undergo an $sp^2 \rightarrow sp^3$ rehybridization and eventually become involved in the MC_3 core. It is interesting to note that this $sp^2 \rightarrow sp^3$ rehybridization takes place at the very last stages of the reaction. Thus, **18a** has essentially retained its olefin and carbene double bonds as well as the sp^2 hybridization around the carbon centers although it has a R(Mo-C_1) distance of 2.40 Å which is only ~.2 Å shorter than the final Mo-C_1 bond length in **14**.

Also, the C_2-C_3 distance at 2.18 Å, **18a**, is much larger than the final C_2-C_3 single bond distance of 1.51 Å in **14**. Thus, the C-C bond formation is seen to lag behind the Mo-C bond formation. Structure **18a** represents a nucleophilic attack of the olefin C_1 carbon **15b** at the metal center. This attack precedes the nucleophilic attack of the carbene C_3 center at the olefinic C_2 carbon, which takes place at the final stages of the reaction with R(C_1-Mo) between 2.40 Å and 2.18 Å. The nucleophilic attack of the olefin C_1 carbon on the metal center involves the ethylene p donor orbital and the a′ metal carbene acceptor orbital, **19a**. The attack is enhanced by an admixture of π^* on olefin which polarizes the donor orbital towards the approaching C_1 carbon.

The subsequent attack by the metal involves donation of charge from the π

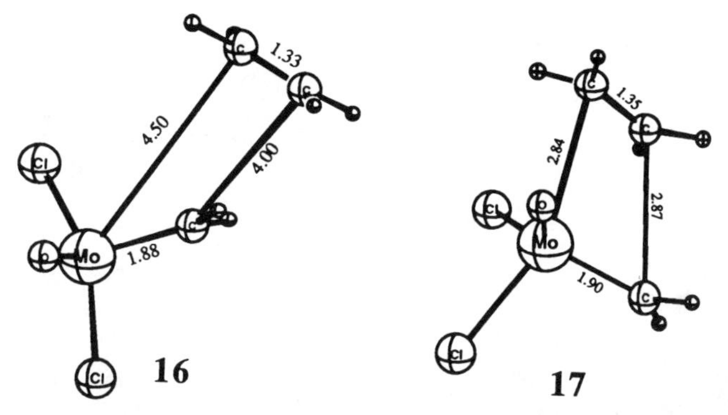

carbene donor orbital, to π^* on olefin which is now polarized towards C_2, **19b**. This donation induces a reduction in the C=C and Mo=C bond orders to one as well as the desired $sp^2 \rightarrow sp^3$ rehybridization. The gain in energy from the $p_{carbene}$ to π^*_{olefin} donation is largely canceled by the energy required (324.8 kJ mol^{-1}) for the $sp^2 \rightarrow sp^3$ rehybridization.

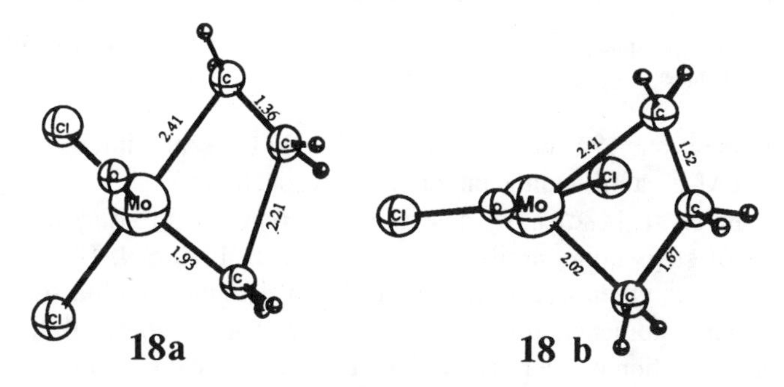

We have seen that the formation of metallacyclobutane in the SP conformation involves transfer of charge to π^*-type orbitals from π donor orbitals, **19a** and **19b**. This transfer is facilitated by the fact that overlaps between π^* and p orbitals are maintained throughout the reaction, **19c** and **19d**. It is also clear from **19c** and **19d** that a concerted approach in which both donor/ac-

C_3 $\pi^*_{carbene}$ + π_{olefin} C_2 C_1 + π^*_{olefin} ⟶ $\pi^*_{carbene}$ +

19a

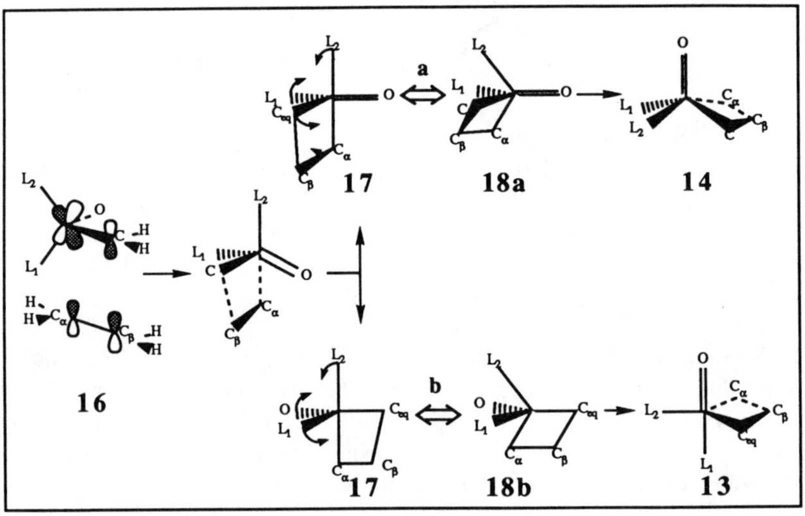

Fig. 5. The two possible Berry pseudorotation routes leading to the formation of SP and TBP metallacyclobutanes.

ceptor interactions **19c** and **19d** are turned on at the same time would lead to C_2-C_3 and M-C_1 antibonding interactions, respectively.

Feldman *et al.* [48c] have described the reaction in equation *20* by a sequence of steps quite similar to those given in Figure *4*. However, their description was given in a somewhat different language with a strong bearing on organic nomenclature. We shall, for the sake of compatibility, reconcile their description of the process with our account. Feldman *et al.* viewed the

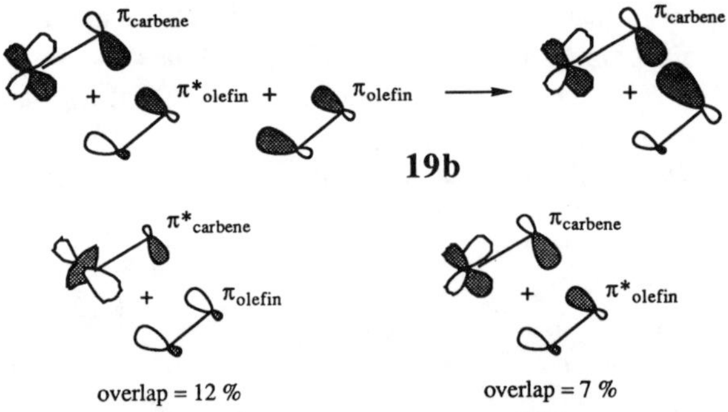

19b

overlap = 12 %

19c

overlap = 7 %

19d

initial stages of the reaction leading to **16**, path **a** of Figure *5*, as an attack of ethylene on the C-L_1-O face of the $L_2Mo(X)CH_2$ carbene. This attack results in the SP olefin complex, **17**. The evolution of the π-complex into the SP intermediate **14** is seen to occur via a Berry-type pseudorotation. According to path a, the spectator oxygen atom, originally in the equatorial plane of the trigonal-bipyramid **17**, adopts an apical position in the *en route* square-pyramid, **18a**. Further transformation leads to the square-pyramidal intermediate structure **14**.

We have now completed our discussion of path **a** in Figure *5* which leads to the formation of the SP metallacycle **14**. We shall next turn to a discussion of the alternative process leading to the TBP metallacycle **13**. This process is indicated in Figure *5* as route **b**. We shall in our study make use of the same $Cl_2Mo(O)C_3H_6$ model system employed for the discussion of **a**. The profile for **b** was traced by starting with $Cl_2Mo(O)C_3H_6$ of the TBP conformation **13** and gradually increasing the Mo-C_1 distance from 2.07 Å in the metallacycle **13** to 5 Å for complete separation of ethylene and the metal carbene Cl_2MoCH_2. The change in the Mo-C_1 distance from 2.07 Å to 5 Å was carried out in 6 steps and in each step all other degrees of freedom were fully optimized. Thus, use has been made of the same Mo-C_1 distance as reaction coordinate for both **a** and **b**. The reaction profile corresponding to path **b** is given in Figure *4b*.

The general geometrical transformation along path **b** is quite simple. Ethylene approaches the metal carbene in the xz plane of **15a** in a way completely analogous to the course taken in path **a**. The only difference between **a** and **b** is that the $Cl_2Mo(O)$ framework hardly changes geometry in **a** whereas the $Cl_2Mo(O)$ moiety in **b** undergoes a drastic deformation as ethylene approaches the carbene molecule. The transformation of the $Cl_2Mo(O)$ fragment in **b** involves a movement of the spectator oxygen in the xy plane of **15a** to a position along the y-direction. At the same time the two chlorines move into the xy plane of **15a** in such a way that Cl_1 ends up trans to oxygen and Cl_2 along the –x direction. We shall now give a more detailed description of path **b** in connection with the profile given in Figure *4b*.

Following path **b** from **17** at R(Mo-C_1) = 2.84 Å to **18b** at R(Mo-C_1) = 2.41 Å leads to an elongation of the olefinic double bond by .16 Å and an increase in the Mo-C_3 bond length by .13 Å. The three carbon centers have in addition undergone an sp^2 to sp^3 rehybridization and the $Cl_2Mo(O)$ fragment has nearly completed the geometrical rearrangement discussed above. The cost in energy of the many deformations is compensated for by the formation of a Mo-C_1 bond as well as a C_2-C_3 carbon bond, **18b**. In total, the energy has gone down by ~4 kJ mol^{-1} in passing from **17** at R(Mo-C_1) = 2.84 Å to **18b** at R(Mo-C_1) = 2.41 Å. After **18b** path **b** proceeds smoothly to **13** at R(Mo-C_1) = 2.07 Å with a total drop in energy of 26 kJ mol^{-1}, Figure *4b*.

There is an important difference between path **a** and **b**. It concerns the timing of the C_2-C_3 carbon bond formation as well as the rehybridization around the carbon centers. In path **a** at R(Mo-C_1) = 2.41 Å, **14**, the C_2-C_3 distance is 2.21 Å with little C-C bond formation and a modest stretch of either the Mo-C_3 carbene distance or the C_1-C_2 olefin bond. Consequently, little rehybridization has taken place around the carbon centers. By contrast at the same value of 2.41 Å, **18b**, for the Mo-C_1 reaction coordinate in **b** an almost full C_2-C_3 carbon bond has formed. In addition, both the Mo-C_3 carbene distance and the C_1-C_2 olefin bond have stretched considerably and the carbon centers have changed to a sp^3 hybridization.

We have previously studied the shape of the donor/acceptor orbitals, **19c** and **19d**, on the metal-carbene fragment of **18a** in path **a** at R(Mo-C_1) = 2.40 Å, and concluded that their shape necessitated a two step attack with Mo-C_1 bond formation preceding C_2-C_3 bond formation. The corresponding donor/acceptor orbitals for the metal-carbene fragment of **18b** in path **b** at R(Mo-C_1) = 2.41 Å are given in **20a** and **20b**, respectively. The donor/acceptor set in **20a** and **20b** are seen to be more prone to concerted Mo-C_1 and C_2-C_3 bond formation. This is clear in **20a** where bonding overlaps can develop along the Mo-C_1 bond and the C_2-C_3 axis. In **20b** the overlap is bonding along the C_2-C_3 vector and nearly zero at the Mo-C_1 terminal. The zero overlap at the Mo-C_1 part arises from the special nodal structure and small amplitude of the metal hybrid.

Our investigation of the reaction between olefin and the $L_2Mo(O)CH_2$ carbene has been restricted to the case where L = Cl. We have not traced the reaction profiles for the other carbene systems with L = OCH_3 and OCF_3. Neither have we discussed the case where the spectator oxygen was replaced by NH. We expect in all cases that the reaction between olefin and the metal carbene will proceed with a modest barrier. Exactly which factors will favor one of the two conformations **11b** and **11c** for the metallacycle will be discussed later. It is further important to point out that SP as well as TBP metallacyclobutane can break up directly to form metalcarbene and olefin. Thus, there is no need to assume [48c] an interconversion of one metallacycle conformation into the other before decomposition.

4.3. SYMMETRY CONSIDERATIONS OF PROCESS FEASIBILITY

Figure *4* indicates that the formation of $L_2Mo(X)C_3H_3$ from ethylene and the metal carbene $L_2Mo(X)CH_2$, equation *20*, has a modest activation barrier. The feasibility of this formally symmetry forbidden $[{}_{\pi}2_s + {}_{\pi}2_s]$ addition reaction

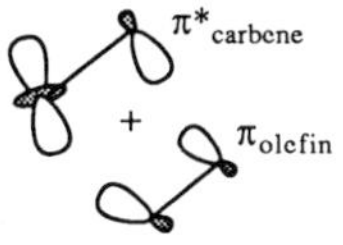

$\pi^*_{carbene}$ + π_{olefin}

overlap = 28 %

20a

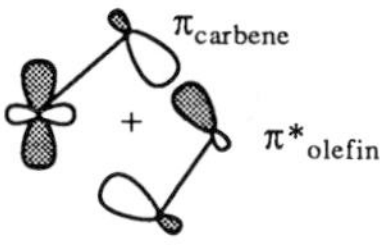

$\pi_{carbene}$ + π^*_{olefin}

overlap = 23 %

20b

is in sharp contrast to the [$_\pi 2_s$ + $_\pi 2_s$] addition reaction between two olefins, equation *21*.

$$(21)\qquad \begin{array}{c} H_2{*}C{=}{*}CH_2 \\ + \\ H_2C{=}CH_2 \end{array} \rightarrow \begin{array}{c} H_2{*}C{-}{*}CH_2 \\ |\quad\ | \\ H_2C{-}CH_2 \end{array} \rightarrow \begin{array}{c} H_2{*}C \\ \| \\ H_2C \end{array} + \begin{array}{c} {*}CH_2 \\ \| \\ CH_2 \end{array}$$

The reaction in equation *21* is symmetry-forbidden by the Woodward-Hoffmann rules [13] and quantitative calculations have revealed an activation barrier of more than 400 kJ mol^{-1}. The barrier stems from the out-of-phase interaction between the occupied π-orbitals on the two approaching ethylenes, $2b_{3u}$ of Figure *6*. This repulsive interaction is increased by the good overlaps between the two π-orbitals as well as their degenerate energies. A transfer of two electrons from the $2b_{3u}$ orbital to the in-phase combination, $1b_{2u}$, between π^*-orbitals on the two approaching ethylenes will eventually eliminate the repulsion. However, the π to π^* charge transfer can not take place before $2b_{3u}$ and $1b_{2u}$ are of the same energy due to the lack of overlap between π and π^* orbitals on opposite ethylenes, hence the energy barrier for the process in equation *21*.

We present a correlation diagram for the formation of metallacyclobutane, **14**, in Figure *7*. The diagram correlates π and σ type orbitals on the reactant side with the orbitals responsible for the four bonds in the ring structure on the product side. The replacement of one ethylene by a metal carbene in the [$_\pi 2_s$ + $_\pi 2_s$] addition reaction, equation *20*, introduces a substantial energy gap between the two occupied π orbitals as π_{olefin} is 4 eV higher in energy than $\pi_{carbene}$. The energy gap will help to reduce the repulsive interaction between the two approaching π-orbitals in the early stages of the organometallic [$_\pi 2_s$ + $_\pi 2_s$] addition reaction of equation *20*.

The final formation of a cyclic structure from two double bonds requires transfer of charge from the occupied π orbitals to the empty π* orbitals. This transfer gave rise to an energy barrier for the formation of cyclobutane due to the lack of π to π* overlaps. The π to π* overlaps are different from zero for the reaction between olefin and metal carbene, **19c–d** and **20a–b**. The non-zero

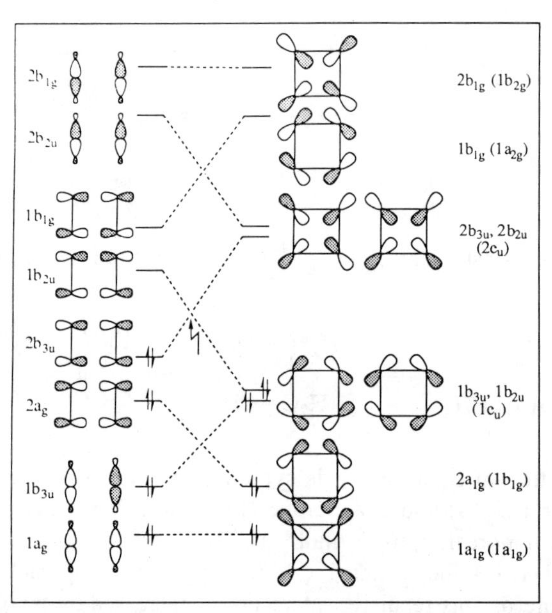

Fig. 6. Correlation diagram for the formation of cyclobutadiene from two molecules of ethylene. Orbital labeling in D_{2h} (and D_{4h}) symmetry point groups.

overlaps allow for a gradual and continous transfer of charge as the cyclic structure is formed, Figure 7. The gradual transfer ensures, in turn, a smooth correlation without a substantial energy barrier. The donation from the π_{olefin} to the carbene acceptor orbital, $\pi^*_{carbene}$, amounts to between .7 e and .8 e depending on the substituents, whereas the back-donation from the $\pi_{carbene}$ to the π^* orbital on C_2H_4 ranges from 1 e to .9 e. In the TBP intermediates we found less donation from the π_{olefin} to the metal centre (between .6 e and .7 e) and consequently less back-donation to the π^*_{olefin} (between .8 e and .9 e). Rappé [50b] *et al.* have provided a rationale for the feasibility of the $[_{\pi}2_s + _{\pi}2_s]$ addition between olefin and metal carbene based on General Valance Bond calculations (GVB). The language in the GVB-theory is somewhat different from that employed in normal orbital arguments. However, the two orbital four-electron repulsions which are of crucial importance in our analysis are also recognized as keyfactors in the GVB-analysis, where they are referred to as Pauli repulsion.

4.4. CONCLUDING REMARKS

The metal carbene $L_2Mo(X)CH_2$ and ethylene can react [48c] along two alternative pathways yielding square-pyramidal or trigonal-bipyramidal

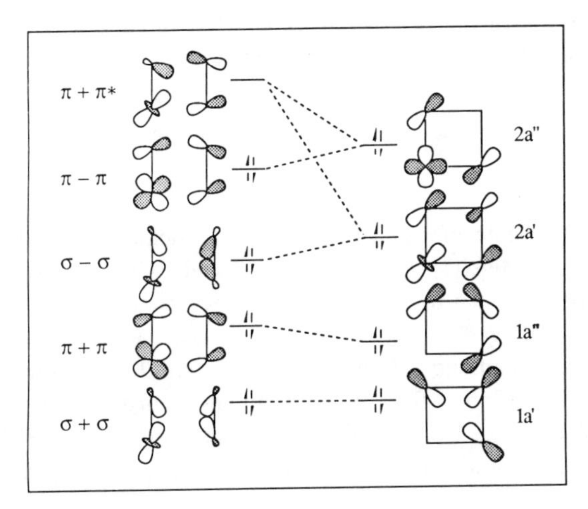

Fig. 7. Correlation diagram for the reaction between carbene and ethylene leading to the formation of the square-pyramidal intermediate **11c** (L = OCH_3, X = NH).

metallacycles, respectively. We have shown that the reactions under investigation shared the same path for the initial stage of the olefin attack onto the carbene molecule, all the way until a π-complex **17** (possibly a transition state) was formed. The reactions can then proceed along one of the two proposed routes. Our work demonstrates that the decomposition of the square-pyramidal intermediate need not occur via a trigonal-pyramidal structure, as once suggested in the literature [48b].

We have calculated the metallacyclic systems to be more stable than the reactants/products. The calculated activation enthalpies for the decomposition of the intermediate metallacycle compare well with the experimental estimates. The reactions in equation *20* were shown to have a small electronic barrier of 10 kJ mol^{-1}.

Based on correlation diagrams, we have contrasted the feasibility of the organometallic $[_\pi 2_s + {}_\pi 2_s]$ processes in equation *20* with the analogous symmetry-forbidden organic $[_\pi 2_s + {}_\pi 2_s]$ reactions between two olefin double bonds (equation *21*). Organic $[_\pi 2_s + {}_\pi 2_s]$ reactions own their high reaction barrier to the lack of interaction between approaching π and π* orbitals as well as large repulsive π-to-π interaction. The repulsive interactions are enhanced further by the close energy of the π orbitals. The repulsive interactions in the organometallic reactions is reduced by a substantial energy gap between the two π orbitals involved. The organometallic reactions are further helped by substantial π to π^* overlaps.

5. Triple bond addition

5.1. INTRODUCTION

High oxidation state tungsten (VI) and molybdenum (VI) alkylidyne complexes such as $(Me_3CO)_3W \equiv CCMe_3$ [53] and $[(CF_3)_2MeCO](Mo \equiv CCH_3)$ [54] have been shown to effectively metathesize acetylenes. The acetylene metathesis is believed to proceed through a dissociative mechanism in which a metallacyclobutadiene [55, 56] intermediate, **21**, is formed as shown in equation *22*. Evidence for this mechanism is provided by the role that metallacyclobutanes, **3**, play in olefin metathesis processes catalysed by analogous alkylidene complexes as shown in equation *3b*.

(22) 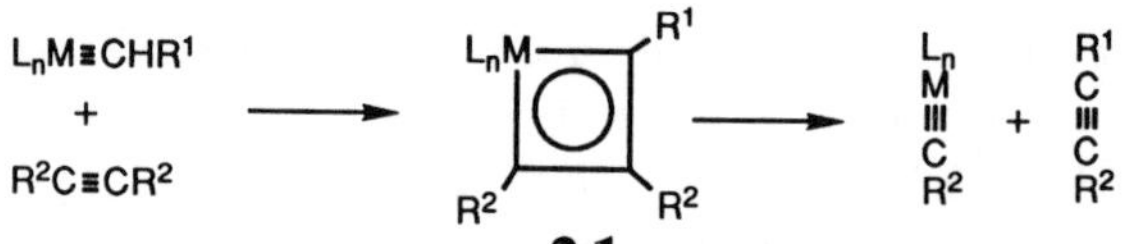 $L_nM{\equiv}CHR^1$ + $R^2C{\equiv}CR^2$ ⟶ [L_nM, R^1, R^2, R^2 metallacycle] **21** ⟶ $L_nM{\equiv}CR^2$ + $R^1C{\equiv}CR^2$

The rate-limiting step in the acetylene metathesis process, equation *22* [56, 57] is expected to be the loss of acetylene from the metallacyclobutadiene. It is thus not surprising that steric bulk of the ancillary ligands on the alkylidyne complex enhance the effectiveness of the catalyst [55, 56, 58, 59].

A potential side reaction in equation *22* is the formation of a metallatetrahedrane complex by tautomerization of the metallacyclobutadiene intermediate **21**. One such example [58, 60] is the formation of $W[\eta^3\text{-}C(CMe_3)C(Me)C(Me)]Cl_3$ from the addition of tetramethylethylenediamine (TMEDA) to the $W[C(CMe_3)C(Me)C(Me)]Cl_3$ metallacyclobutadiene as shown in equation *23* [58].

(23) 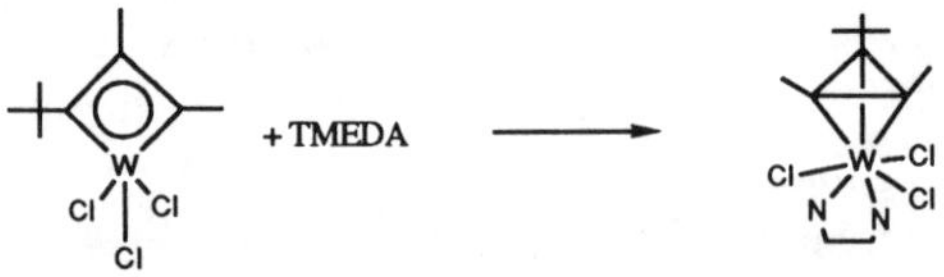 [W, Cl, Cl, Cl metallacycle] + TMEDA ⟶ [Cl—W—Cl, Cl, N N metallatetrahedrane]

Metallatetrahedranes such as $W[\eta^3\text{-}C_3\text{-}(CMe_3)Et_2](O_2CMe)_3$ can also form from the direct addition of acetylene to a metal alkylidyne complex [61]. The formation of stable metallatetrahedranes might inhibit the metathesis of acetylenes [62].

Both the metallacyclobutadiene and the metallatetrahedrane have been examined theoretically. Anslyn *et al.* [63a] and Bursten [63b] have studied the nature of the bonding in the metallacyclobutadiene and metallatetrahedrane. Jemmis and Hoffmann [63c] have treated the isomerization of the metallacyclobutadiene and the metallatetrahedrane. A detailed study of the formation and decomposition of both the metallacyclobutadiene and the metallatetrahedrane have not been presented previously.

We shall first study the formation and decomposition of the metallacyclobutadiene, $Cl_3MoC_3H_3$. Then, in Section 5.4, a similar study will be given for the formation of the corresponding metallatetrahedrane. Our analysis of the reaction pathways will be based on orbital symmetry considerations and quantitative Density Functional calculations. Although most acetylene metathesis systems involve tungsten carbyne catalysts with alkoxide ligands, molybdenum was used in place of tungsten and the alkoxide ligands were substituted by chlorine atoms. Both modifications were introduced in order to reduce the computational cost. There is usually a close correspondence between the second and third transition series metals of the same triad, especially for tungsten and molybdenum complexes, and it was felt that the substitution would not significantly effect the results. In fact, many molybdenum carbyne [54, 64] metallacyclobutadiene [54] and metallatetrahedrane [63, 65] systems have emerged including several molybdenum metathesis systems [54, 66].

5.2. FORMATION OF METALLACYCLOBUTADIENE

A detailed study on the formation of the metallacyclobutadiene was carried out in a postulated least motion pathway as shown in equation *24*.

(24)

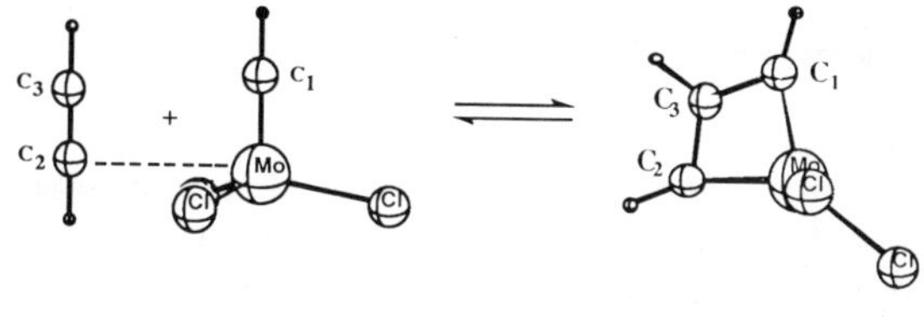

The molybdenum carbyne complex Cl_3MoCH, **22a**, was brought together with the acetylene molecule such that the three carbon atoms and the metal centre remained coplanar throughout the course of the reaction. The Mo-C_2 distance was used as a reaction coordinate and varied between 5.29 Å, a distance where the two molecules are virtually non-interacting, to 1.92 Å, the distance of the Mo-C_2 bond length in the final metallacyclobutadiene, **23**.

A total of eight steps along the reaction profile were calculated by fixing the Mo-C_2 distance and optimizing all other degrees of freedom within a C_s symmetry constraint. The energy profile for this reaction is shown in Figure *8*. The labels on the plot will be discussed shortly.

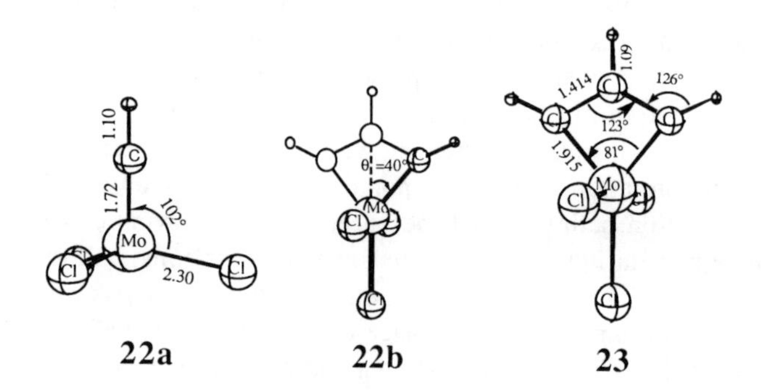

The reaction profile reveals that the metallacyclobutadiene is 71 kJ mol^{-1} more stable than the two reacting species and that there is a small 10 kJ mol^{-1} reaction barrier with a maximum roughly occuring at a Mo-C_2 distance of 3.18 Å.

The formation of metallacyclobutadiene is very similar to the symmetry forbidden formation of cyclobutadiene from acetylene except that in the organometallic reaction one carbon is replaced by a metal centre. We shall in the following try to understand how the organometallic reaction in equation *25* is able to circumvent the sizable barrier encountered in the direct formation of cyclobutadiene from two acetylene molecules in a least motion pathway, equation *25*.

(25) $\quad + \quad \xrightarrow{\Delta}$

Similar to the reaction in equation *21*, the reaction in equation *25* correlates an occupied π-orbital, $2b_{3u}$, on the reactant side with an empty $2e_u$ orbital on the product side (Figure *6*) as well as an empty π^*-orbital, $1b_{2u}$, on the reactant side with an occupied $1b_{2u}$ orbital on the product side. The net result is a transfer of charge (2e) from π to π^*. This transfer of charge unfortunately cannot take place gradually since the π and π^* type orbitals are of different symmetries and thus non interacting. The charge transfer must instead await the point where $2b_{3u}$ has been destabilized sufficiently to be of the same energy as $1b_{2u}$. However the destabilization of the fully

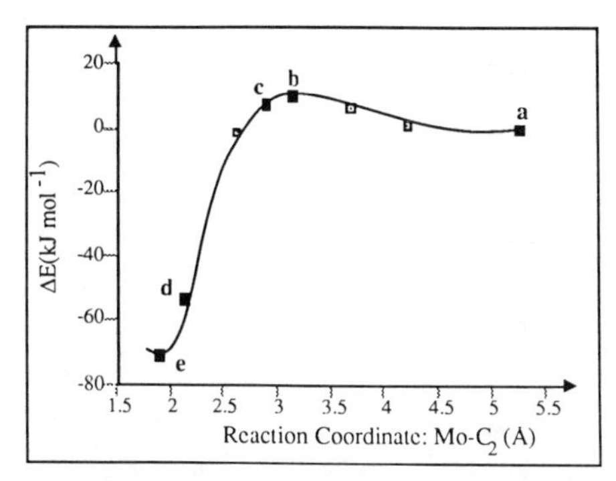

Fig. 8. Energy profile for the formation of the metallacyclobutadiene.

occupied $2b_{3u}$, (π-π), orbital translates into a sizable activation barrier.

Figure *9* provides a correlation diagram for the formation of σ bonds of the metallacyclobutadiene according to the least motion pathway of equation *24*. This pathway has a preserved C_s symmetry with the C_s plane being the plane of the ring in the resulting metallacycle. In this approach all orbitals in the σ-framework are of a′ symmetry and the orbitals are labelled accordingly. Orbitals 1a′ and 2a′ of the reactants are the plus and minus combinations, respectively, of the σ orbitals of the acetylene and the metal carbyne. Similarly the 3a′ and 4a′ orbitals of the reactants are the plus and minus combinations of the π orbitals of the acetylene and carbyne. Figure *9* shows that these four occupied orbitals of the reactants correlate smoothly to four product orbitals which form the sigma bonding framework of the metallacycle. This is in contrast to the organic reaction in which the (π-π) orbital of the reacting acetylenes interacted repulsively and correlated to an excited state of the product, Figure *6*.

In the organometallic reaction the repulsive interaction between the two π orbitals, 4a′ of Figure *9*, is reduced by two factors. First, the π-orbital on acetylene is 5 eV lower in energy than the π-orbital on the metal carbyne. This gap will tend to reduce the repulsive interaction in 4a′ during the initial stages of the process in equation *24*. Secondly, π and π^* orbitals can overlap throughout the process in equation *24* thus allowing for a gradual transfer of charge from π to π*. Not only do the π and π^* orbitals interact with one another in the organometallic case, but these interactions are actually stronger than the π + π interactions. This can be illustrated by examining the overlaps between such orbitals as shown in **24**.

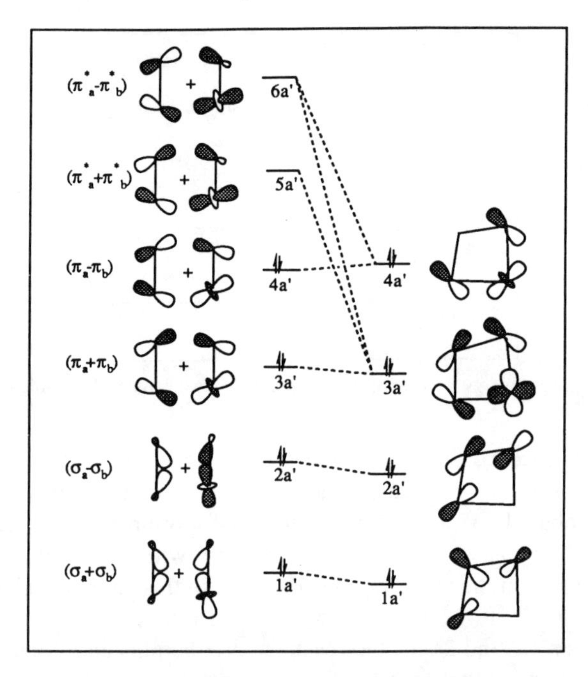

Fig. 9. Correlation diagram for the formation of the σ-framework of the metallacyclobutadiene.

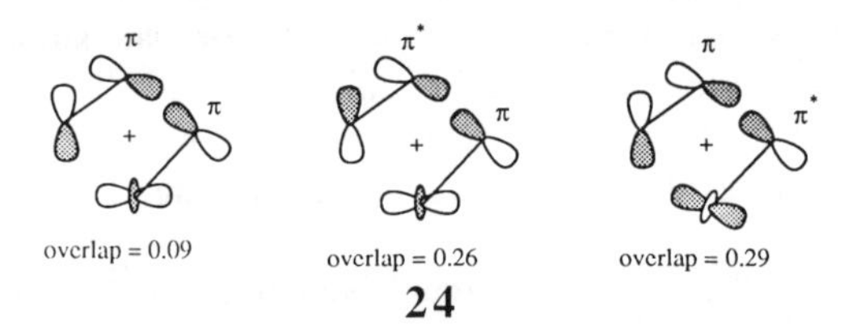

The rather large overlaps between the π and π^* combinations results from the odd π and π^* orbitals formed by the molybdenum carbyne, **22b**, when it is distorted into the geometry taken up in the metallacyclobutadiene. The hybridization of the d orbitals of the molybdenum result in a π^* orbital, for example, resembling a π orbital to an incoming acetylene molecule, **24**. The protruding lobes of a π^* orbital on carbyne are of the same sign and resemble a π-type orbital to an incoming acetylene allowing for the large overlaps between π and π^* orbitals. In a similar manner, a carbyne π orbital has protruding lobes of opposite sign and resembles a π^* orbital to an incoming acetylene. During the course of the reaction, the non-zero overlap between π and π^* orbitals allows for donation of electron density from the π orbital of the acetylene into the empty π^* orbital of the carbyne and back donation from the π orbital of the carbyne into the empty π^* orbital of the acetylene, **25**. It is primarily this feature that distinguishes the organometallic reaction from the organic reaction and allows for stabilization of orbitals through donor-acceptor type interactions between π and π^* orbitals.

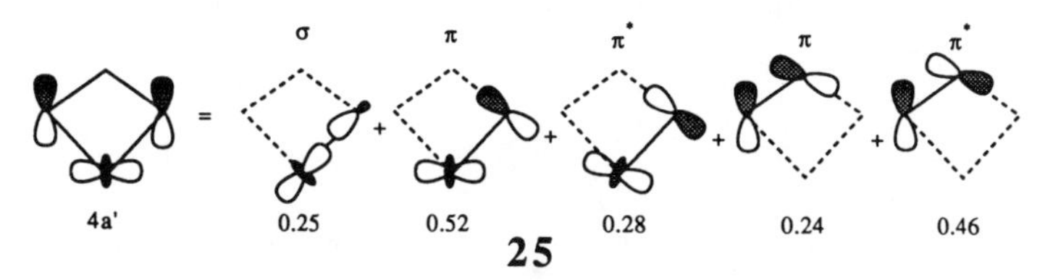

25

A correlation diagram for the change in the π system perpendicular to the C_s plane is given in Figure *10*. The diagram in Figure *10* resembles in many ways that drawn [67] for the formation of the π-system in cyclobutadiene.

The most significant difference is that the 2a″ and 3a″ orbitals are degenerate and nonbonding in the square conformation of cyclobutadiene. This degeneracy will result in a Jahn-Teller distortion into a conformation with alternate bond lengths. In metallacyclobutadiene 2a″ is bonding and of lower energy than 3a″. The bonding character in 2a″ is an attribute of the d-type

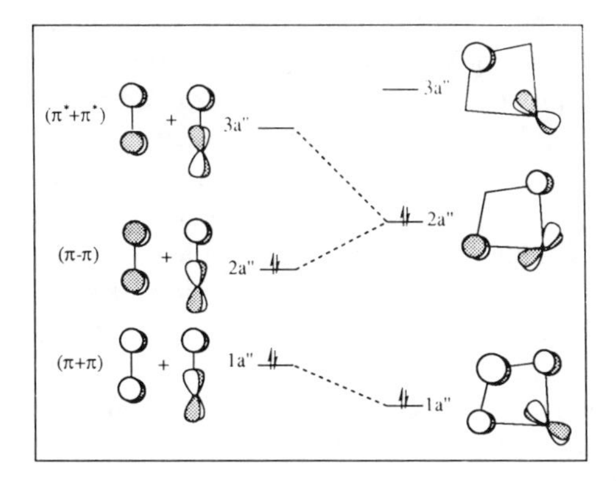

Fig. 10. Correlation diagram for the formation of the π-framework of the metallacyclobutadiene.

orbitals on the metal centre and enforce a conformation, **23**, with respectively equidistant C_a-C_b and M-C_a distances. Most metallacyclobutadienes have a planar, delocalized π system similar to that of **23**. However, some metallacyclobutadienes such as $CpCl_3WC(Ph)C(CMe_3)C(Ph)]$, **26**, have alternate double and single bonds despite being symmetrically substituted [68]. This type of distortion that is observed in **26** can result from mixing of the 2a″ orbital with the 3a″ LUMO as shown in **27**. This results in a partial localization of the π system and an alternate double and single bond length pattern. The localization of the π bonding also allows for the ring puckering which is observed in these metallacycles.

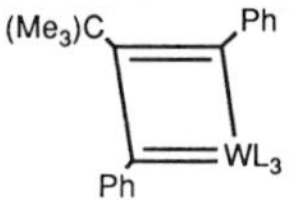

$(Me_3)C$ Ph WL_3 Ph

26 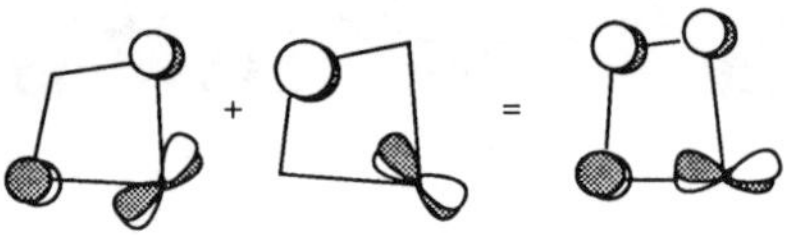 **27**

5.3. A STUDY OF THE PATH LEADING TO METALLACYCLOBUTADIENE

Up to this point, the discussion of the reaction in equation *24* has been concentrated on why the formation of metallocyclobutadiene is feasible with a modest barrier whereas the corresponding reaction between two acetylene molecules is "symmetry forbidden" with a very high activation barrier. We shall now turn to a more detailed account of the reaction in equation *24* based on the profile presented in Figure *8*. The labels **a–d** on the energy profile of Figure *8* refer to structures **28a–28d**, and point **e** corresponds to the strucure **23**. The starting structure **28a** has a Mo-C_l distance of 5.29 Å. At this point of the process there is virtually no interaction between the reactants and both acetylene and carbyne are in their free undistorted states. As the Mo-C_l distance is decreased from 5.29 Å the energy steadily rises until it reaches a maximum labelled **b** which is a total of 10 kJ mol^{-1} higher in energy than point **a**. This maximum occurs at a Mo-C_l distance of 3.18 Å and **28b** can be considered an approximate tranisition state (no attempt has been made to verify whether or not it has a single normal mode with an imaginary frequency).

An examination of the molecular orbitals of the approximate transition state **28b** reveals that except for one, all of the σ, π, and π^* orbitals discussed earlier are localized on either the carbyne or acetylene fragments as they would be in the free species. The one exception is the orbital, **29a**, which correlates to the

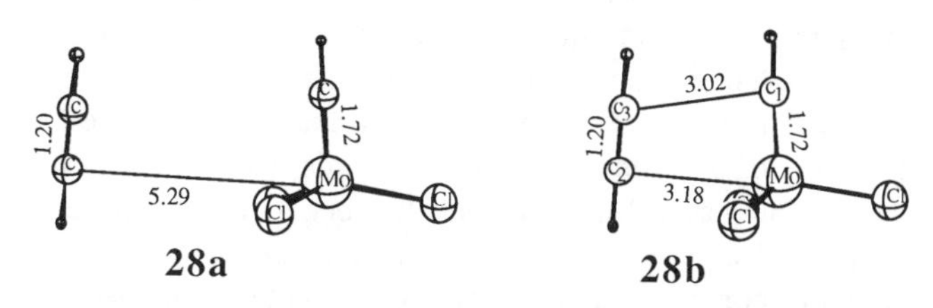

carbyne π orbital on the reactant side, 4a′ in Figure *9*. This orbital, although polarized towards the carbyne fragment, has a significant contribution from the π orbital of acetylene which mixes in an antibonding fashion, **29a**.

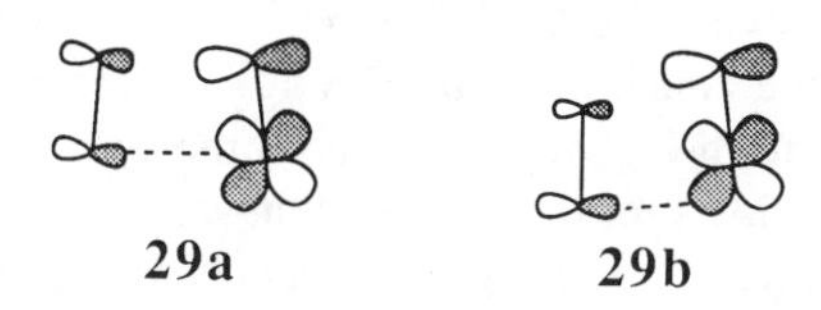

In moving away from the transition state toward the final product, this π-π destabilizing interaction, **29a**, is relieved by electron donation into the π^* orbital of acetylene. The π^* acetylene orbital mix in by decreasing the orbital lobe on C_3 which in turn results in a decreased π-π antibonding interaction between the two fragments, **29b**. This effect is readily apparent as early on as point **c** on the reaction profile corresponding to **28c** where the Mo-C_2 distance is 2.91 Å. The mixing or the π^* orbital which polarizes the π-π orbital towards C_2 on the acetylene fragment also facilitates a nucleophilic attack of C_2 on the molybdenum centre. It is evident from **29b** corresponding to the structure **28c** that the acetylene fragment has swung around toward the chlorine ligands, allowing the orbital lobe on C_2 to interact in a stablizing fashion with another lobe on the molybdenum d_{xy} orbital. This type of stabilization involving the metal d orbital clearly cannot happen in the organic reaction between two acetylenes to form cyclobutadiene.

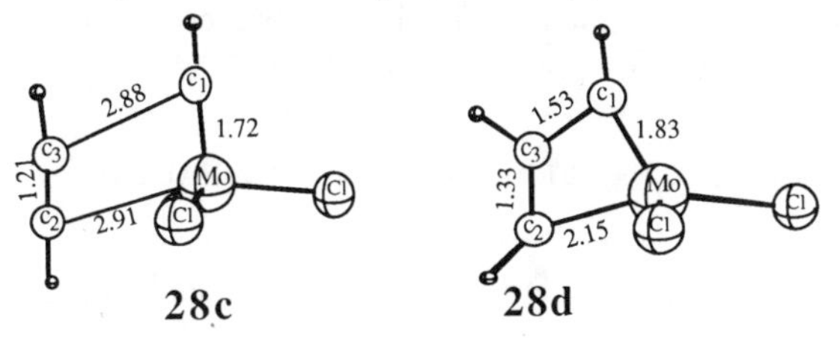

The structure **28c** is a weakly bonded Mo-C_2 complex, with the other σ and π bonds remaining completely localized on one of the two fragments. The Mo-C_1 bond length is unchanged from the free carbyne, while the mixing of the π^* orbital from the acetylene fragment has elongated the C_2-C_3 bond only slightly by .01 Å. The two chlorine ligands that were initially directed toward the incoming acetylene fragment have begun to swing around, moving toward their final axial positions in the metallacyclobutadiene. Following the formation of complex **28c** there is a steep decrease in energy. With this steep descent there is further mixing of the π^* of the acetylene fragment allowing for a stronger Mo-C_2 bond to form. This nucleophillic attack of the molybdenum is then followed by a polarization of the carbyne π orbital away from the metal centre towards C_1. This is achieved by electron donation into the molybdenum carbyne π^* orbital. The polarization towards C_1 then facilitates the nucleophillic attack of C_3 by C_1 allowing for the ring closure. The electron donation into the π^* orbitals of both the carbyne fragment and the acetylene fragment allows for the three carbon atoms to undergo the sp to sp^2 rehybridization required to form the metallacyclobutadiene.

It is interesting to note that the ring closure and rehybridization do not occur until the late stages of the reaction as shown in **28d**, which has Mo-C_2 bond distance of 2.15 Å (only .25 Å from the final bond distance in the metallacyclobutadiene). The donation and back donation into the π^* orbitals of the metal carbyne and acetylene fragments is evident by the Mo-C_1 and C_2-C_3 bond distances which are roughly halfway between that found in the reactants and the final metallacycle. The structure of **28d** illustrates that the hydrogen substituents have bent back towards their final positions in the metallacyclobutadiene showing the rehybridization that is taking place. Also, the two chlorine atoms have also assumed their axial positions in the final metallacyclobutadiene.

The picture developed here with an initial nucleophilic attack of the metal centre followed by a subsequent nucleophilic attack of C_3 by C_1 is consistent with experimental results where the effectiveness of various $X_3W \equiv C$-Y metal carbynes as metathesis catalyst was studied [57]. From these studies where the X ligand and the Y substituent were varied it was concluded that the $W \equiv C$ bond behaved as if it were polarized, $W(\delta^+) \equiv C(\delta^-)$.

5.4. FORMATION OF METALLATETRAHEDRANE FROM METAL CARBYNE AND ACETYLENE

Metallatetrahedranes are known to form metallacyclobutadienes upon addition

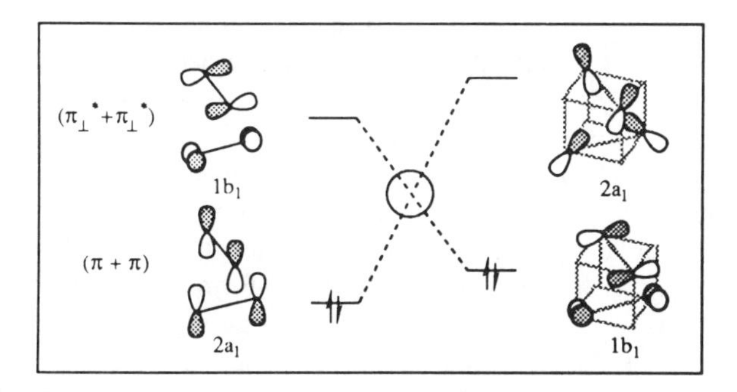

Fig. 11. Correlation diagram for the formation of tetrahedrane.

of donor ligands and from direct addition of acetylene to metal alkylidynes [58–60]. In the following, we shall study the direct formation of the metallatetrahedrane $Cl_3Mo(\eta^3\text{-}C_3H_3)$, **30**, in a least motion approach according to equation *26*.

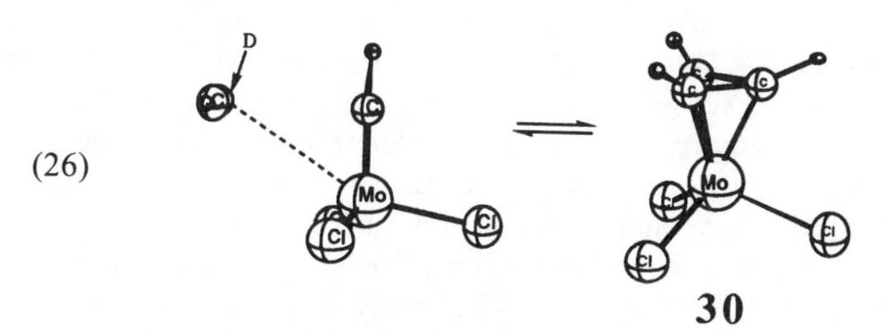

We shall compare this reaction to the least motion perpendicular approach of two acetylene molecules to form tetrahedrane, as in equation *27*.

(27)

A correlation diagram for the process in equation *27* is given in Figure *11*. The correlation diagram reveals that the occupied $2a_1(\pi + \pi)$ orbital of the reactants correlates to an empty product orbital. Similarly, the empty $1b_1(\pi_{\perp}^{*} + \pi_{\perp}^{*})$ orbital of the reactant correlates to an occupied product orbital, thus making the reaction symmetry forbidden. The skew approach of the acetylenes affords the same basic nodal structure of the orbitals and would also be symmetry forbidden.

A correlation diagram for the formation of metallatetrahedrane according to equation *26* is given in Figure *12*. The correlation diagram reveals that the direct formation of metallatetrahedrane by the perpendicular approach of the acetylene is forbidden by orbital symmetry just as the formation of the tetrahedrane is forbidden. The occupied 5a′ orbital which corresponds to a carbyne π bonding orbital correlates to an unoccupied σ^* orbital of the metallatetrahedrane. The empty acetylene π^* orbital of a″ symmetry correlates to an occupied Mo-C σ bonding orbital of the metallatetrahedrane. Consequently, for the metallatetrahedrane to form, the electron pair in the 5a′ orbital of the reactants must at some point during the course of the reaction jump into the 2a″ π^* orbital of acetylene. Since a change in occupancy of the orbitals of different symmetry can only occur when the two orbitals are of equal energy, the formation of the metallatetrahedrane by the perpendicular approach is expected to have at least a modest energy barrier. The magnitude of the energy barrier is dependent on where the crossing occurs. Thus, unlike the formation of the metallacyclobutadiene by the least motion pathway the formation of the tetrahedrane is formally symmetry forbidden. We have determined an energy barrier for the symmetry forbidden formation of the metallatetrahedrane as shown in equation *26*. In this set of calculations the metallatetrahedrane is formed by the perpendicular approach of the reactants such that there is a preserved plane of symmetry (plane of the paper in equation *26*). The distance between the molybdenum atom and the point D, the point colinear with and exactly midway between the two carbons of the acetylene molecule, is used as a reaction coordinate (this is shown in equation *26* as a dotted line). In this way, the molybdenum and the point D remain on the preserved C_s plane. For all steps, the Mo-D distance was fixed, while all other degrees of freedom were optimized within the C_s symmetry constraints.

To represent the crossing that must occur in this forbidden reaction, two sets of calculations were performed, one in the electron configuration of the free species and the other in the electron configuration of the metallatetrahedrane. More specifically, for one set of calculations the metallatetrahedrane was formed by fixing the electron configuration to that of the free species (left side of Figure *12*). In this way, the occupied 5a′ orbital of the free species is significantly raised in energy as the Mo-D distance is decreased. In the other set of calculations the metallatetrahedrane was decomposed by fixing its electron configuration to that of the right side of Figure *12*. This caused 2a″ orbital to increase in energy as the Mo-D distance was increased.

The energy of the 5a′ orbital of the free species and the 2a″ orbital of the metallatetrahedrane as the Mo-D distance is varied is shown in Figure *13a*. The crossing is circled and occurs at a Mo-D distance of 2.55 Å. The total

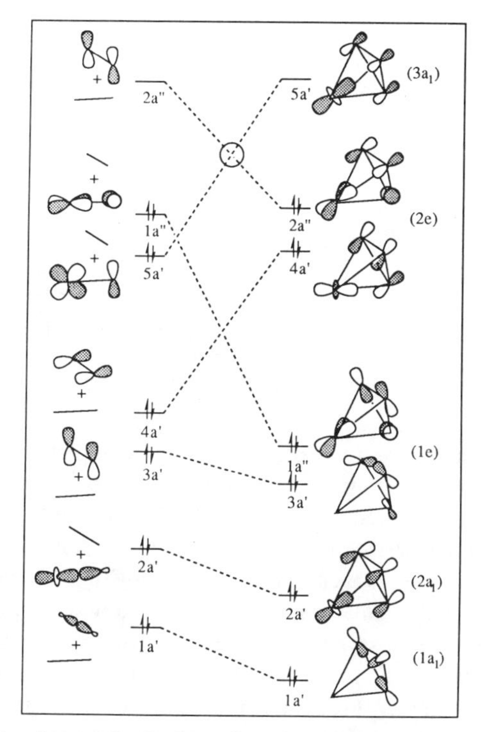

Fig. 12. Correlation diagram for the formation of metallatetrahedrane from acetylene and metal carbyne.

energy of the complex for both electron configurations relative to the free species is shown in Figure *13b*. The solid line represents the energy profile for the forbidden reaction. The energy profile traces the total energy of one configuration until the crossing distance of 2.55 Å, at which point it traces the total energy of the other electron configuration.

Figure *13b* shows that there is only a modest 40 kJ mol^{-1} electronic barrier for the direct formation of the molybdenum tetrahedrane from the acetylene and molybdenum carbyne. Furthermore, although it is not shown in Figure *13b*, the metallatetrahedrane is 122 kJ mol^{-1} more stable than the free acetylene and molybdenum carbyne.

5.3. CONCLUDING REMARKS

A density functional study of acetylene metathesis catalyzed by high oxida-

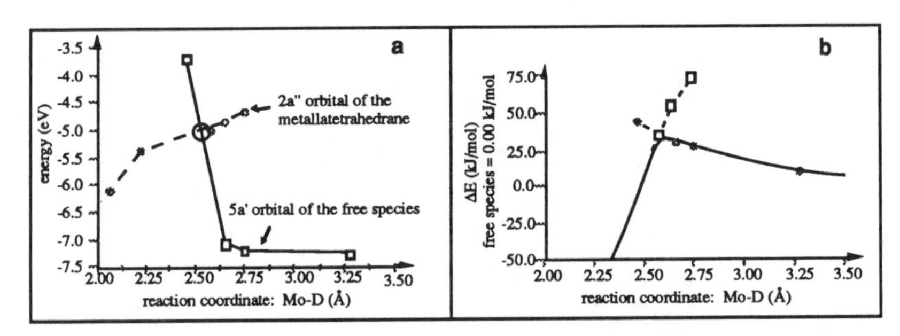

Fig. 13. (a) Correlation of the 5a′(□) and 2a″(O) orbitals for the formation of the metallatetrahedrane. (b) Energy profile for the formation of metallatetrahedrane; O – Total energy of the free species, □ – Total energy of the metallatetrahedrane.

tion state molybdenum and tungsten alkylidyne complexes was undertaken. In the first part of the study, the feasibility of a dissociative mechanism of metathesis was investigated where a metallacyclobutadiene acts as an intermediate. The geometries, bonding, and energetics of the formation and decomposition of the metallacyclobutadiene, $Cl_3MoC_3H_3$, **23**, were determined. The formation of the metallacyclobutadiene was found to be fully symmetry allowed and possessed a small 10 kJ mol^{-1} electronic barrier. The formation of the metallacyclobutadiene has a small activation barrier since the π and π^* type orbitals can interact throughout the course of the reaction. It was also found that the formation of the metallacyclobutadiene is initiated by the nucleophilic attack of the acetylene on the metal centre. The metallacyclobutadiene was determined to be 71 kJ mol^{-1} more stable than the acetylene and molybdenum carbyne reactants revealing that the rate determining step is the decomposition of the metallacyclobutadiene.

In the second part of this study we investigated the formation of metallatetrahedrane $Cl_3Mo(\eta^3 - C_3H_3)$, **30**, by the direct addition of the metathesis reactants. The direct formation of metallatetrahedrane was found to be forbidden by orbital symmetry and was determined to possess a 40 kJ mol^{-1} electronic barrier. The metallatetrahedrane, **30**, was also found to be 52 kJ mol^{-1} more stable than its corresponding metallacyclobutadiene, **23**, and 122 kJ mol^{-1} more stable than the free acetylene and molybdenum carbyne. The modest 40 kJ mol^{-1} reaction barrier for the direct formation of the metallatetrahedrane reveals that the metallatetrahedrane is energetically accessible. It would be interesting to combine molecular mechanics and density functional calculations in an attempt to determine whether more bulky ligands on the metal centre can influence the relative stability of metallacyclobutadiene and metallatetrahedrane.

6. Conclusions

With Density Functional based method as a tool, we have examined a series of organometallic $[2_s + 2_s]$ reactions of major industrial interest. The focus has been on the mechanisms allowing for the feasibility of these transition metal catalysed reactions and contrasting them with their forbidden, purely organic counterparts.

For early transition metals we have studied the $[_\sigma 2_s + {}_\sigma 2_s]$ single bond metathesis reactions. The systems included reactions between Cp_2Sc-R (R = H, CH_3) and H-H as well as C-H bonds in CH_4, C_2H_4 and C_2H_2. The rather low energy barriers resulted from the mixing in of the empty d orbitals on scandium into the high-lying, non-bonding orbitals in the four-centre transition states. A pool of empty s, p and d orbitals on scandium was seen to continuously rehybridize the metal component of the reacting Cp_2Sc-R complexes providing the stabilization of the transition states at all times. It was also found, that the presence of hydrogen atoms (with their spherical 1s orbitals) in the ring system stabilized the transition state structures and lowered the energy barriers. The calculated energy barriers were seen to increase with the number of carbon atoms involved in the cyclic interaction and were related to the inability of directional σ-orbitals on carbon atoms to provide efficient overlaps with the adjacent atoms. We have also examined the possibility of competing reactions, ie. insertion of olefins and acetylenes into the Sc-H and Sc-C(H_3) bonds. It was concluded, on purely thermodynamic grounds and in agreement with the experimental evidence, that the insertion of olefins was favoured over the alkenylic C-H activation. However, in contrast to the experiment, the insertion of acetylenes was found to be preferred over the alkynylic C-H activation.

For middle transition metals we have studied the formally $[_\pi 2_s + {}_\pi 2_s]$ double and triple bond metathesis reactions catalysed respectively by molybdenum carbene and carbyne complexes. Both processes were shown to be symmetry allowed with small (~10 kJ mol^{-1}) energy barriers proceeding the formation of the four-centre intermediates. Such factors as the non-zero $\pi_{carbene}$-π^*_{olefin}, $\pi^*_{carbene}$-π_{olefin}, $\pi_{carbyne}$-$\pi^*_{acetylene}$ and $\pi^*_{carbyne}$-$\pi_{acetylene}$ overlaps were responsible for the lowering of the energy barriers. The formation of the metallatetrahedrane from the direct approach of metal carbyne and acetylene was examined as a potential side reaction to the acetylene metathesis process. Although it was shown to be symmetry forbidden, a modest electronic barrier of ~40 kJ mol^{-1} indicates a high possibility of its occurrence.

This investigation was supported by the Natural Sciences and Engineering Research Council of Canada (NSERC). We would also like to acknowledge access to the IBM-6000 installations at the University of Calgary.

Elzbieta Folga, Tom Woo and Tom Ziegler
Department of Chemistry,
University of Calgary, Calgary, Alberta T2N 1N4, Canada

References

1. P. Cossee, J. Catal., **3**, 80 (1964)
2. J.-L. Hérisson and Y. Chauvin, Makromol. Chem., **141**, 161 (1979)
3. (a) P.L. Watson and G.W. Parshall, Acc. Chem. Res., **18**, 51 (1985); (b) P.L. Watson, J. Chem. Soc., Chem. Commun., 276 (1983); (c) P.L. Watson, J. Am. Chem. Soc., **105**, 6491 (1983)
4. (a) A.K. Rappé and T.H. Upton, J. Am. Chem. Soc., **114**, 7507 (1992); (b) A.K. Rappé, Organometallics, **9**, 466 (1990)
5. H.-K. Kuribayashi, N. Koga and K. Morokuma, J. Am. Chem. Soc., **114**, 2359 (1992)
6. N. Koga and K. Morokuma, Chem. Rev., **91**, 823 (1991)
7. C.A. Jolly and D.S. Marynick, J. Am. Chem. Soc., **111**, 7968 (1989)
8. H. Rabaâ, J.-Y. Saillard and R. Hoffmann, J. Am. Chem. Soc., **108**, 4327 (1986)
9. H. Fujimoto, T. Yamasaki, H. Mizutani and N. Koga, J. Am. Chem. Soc., **107**, 6157 (1985)
10. (a) T. Ziegler, E. Folga and A. Berces, J. Am. Chem. Soc., **115**, 636 (1993); (b) E. Folga and T. Ziegler, Organometallics, **12**, 325 (1993); (c) T.K. Woo, E. Folga and T. Ziegler, Organometallics, **12**, 1289 (1993)
11. M.L. Steigerwald and W.A. Goddard, J. Am. Chem. Soc., 106, **308** (1984)
12. (a) A.D. Becke, Phys. Rev., **A 38**, 2398 (1988); (b) J. Perdew, Phys. Rev., **B33**, 8822 (1986); ibid, (1986), **B34**, 7046
13. R.B. Woodward and R Hoffmann, The Conservation of Orbital Symmetry; Verlag Chemie, Weinheim (1970)
14. E.J. Baerends, D.E. Ellis and P. Ros, Chem.Phys., **2**, 41 (1973)
15. E.J. Baerends, Ph.D. Thesis, Vrije Universiteit, Amsterdam (1975)
16. W. Ravenek, *A*lgorithms and Applications on Vector and Parallel Computers, H.J.J. te Riele, Th. J. Dekker and H.A. van de Vorst, Eds., Elsevier, Amsterdam (1987)
17. (a) P.M. Boerrigter, G. teVelde and E.J. Baerends, Int. J. Quantum Chem., **33**, 87(1988); (b) G. teVelde and E.J. Baerends, J. Comp. Phys., **99**, 84 (1992)
18. L. Versluis and T. Ziegler, J. Chem. Phys., **88**, 322 (1988)
19. (a) G.J. Snijders, E.J. Baerends and P. Vernooijs, At. Nucl. Data. Tables **26**, 483 (1982); (b) P. Vernooijs, G.J. Snijders and E.J. Baerends, *S*later Type Basis Functions for the whole Periodic

System, Internal report, Free University of Amsterdam, The Netherlands (1981)
20. J. Krijn and E.J. Baerends, *F*it functions in the HFS-method, Internal Report (in Dutch), Free University of Amsterdam, The Netherlands (1984)
21. S.H. Vosko, L. Wilk and M. Nusair, Can. J. Phys, **58**, 1200 (1990)
22. A.D. Becke, Phys. Rev., **A38**, 2398 (1988)
23. (a) T. Ziegler, J. Pure. Appl. Chem., **63**, 873 (1991); (b) T. Ziegler and L. Versluis, ACS Adv. in Chemistry Series, **230**, 75 (1992); (c) T. Ziegler and T. Tschinke, ACS Symposium Series, **428**, 277 (1990); (d) T. Ziegler, J.G. Snijders and E.J. Baerends, ACS Symposium Series **383**, 322 (1989); (e) T. Ziegler, Chem. Rev., **91**, 651 (1991); (f) T. Ziegler, V. Tschinke and L. Versluis, NATO ASI Series, **C176**, 189 (1986)
24. (a) R.G. Parr and W. Yang, *D*ensity-Functional Theory of Atoms and Molecules, Oxford University Press, New York (1989); (b) E.S. Kryachko and E.V. Ludena, *D*ensity Functional Theory of Many Electron Systems, Kluwer Press, Dordrecht (1991)
25. A.D. Becke, Int. J. Quantum Chem., **S23**, 599 (1989)
26. (a) T. Ziegler, V. Tschinke, L. Versluis, E.J. Baerends and W. Ravenek, Polyhedron, **7**, 1625 (1988)
27. L. Versluis and T. Ziegler, J. Chem. Phys., **88**, 322 (1988)
28. R. Fournier, J. Andzelm and D.R. Salahub, J. Chem. Phys., **90**, 6371 (1989)
29. L. Fan, L. Versluis, T. Ziegler, E.J. Baerends and W. Ravenek, Int. J. Quantum Chem., **S22**, 173 (1988)
30. (a) W. Bieger, G. Seifert, H. Eschrig and G. Grossman, Chem. Phys. Lett., **115**, 275 (1985); (b) D.A. Freier, R.F. Fenske and Y. Xiao-Zeng, J. Chem. Phys., **83**, 3526 (1985) (c) V.G. Malkin and Z. Zhidomirov, Zh. Strukt. Khim., **29**, 32 (1988)
31. A.J. van der Est, P.B. Barker, E.E. Burnell, C.A. de Lange and J.G. Snijders, Mol. Phys., **56**, 1 (1985)
32. D.A. Case, Annu. Rev. Phys. Chem., **33**, 151 (1982)
33. (a) L. Noodleman and J.G. Norman, J. Chem. Phys., **70**, 4903 (1979); (b) L. Noodleman, J. Chem. Phys., **74**, 5737 (1981); (c) L. Noodleman and E.J. Baerends, J. Am. Chem. Soc., **106**, 2316 (1984); (d) L. Noodleman, J.G. Norman, J.H. Osborne, A. Aizman and D.A. Case, J. Am. Chem. Soc., **107**, 3418 (1985)
34. L. Fan and T. Ziegler, J. Chem. Phys., **92**, 3645 (1990)
35. M. Trsic, T. Ziegler and W.G. Laidlaw, Chem. Phys., **15**, 383 (1976)
36. M. Cook and M. Karplus, J. Phys. Chem., **91**, 31 (1987)
37. V. Tschinke, and T. Ziegler, J. Chem. Phys., **93**, 8051 (1990)
38. G.J. Kubas, Acc. Chem. Res., **21**, 120 (1988)
39. (a) A.H. Janowicz and R.G. Bergman, J. Am. Chem. Soc., **104**, 352 (1982); (b) J.K. Hoyano and W.A.G. Graham, J. Am. Chem. Soc., **104**, 3723 (1982); (c) W.D. Jones and F.J. Feher, J. Am. Chem. Soc., **106**, 1650 (1984); (d) W.D. Jones and F.J. Feher, Acc. Chem. Res., **22**, 91 (1989); (e) I.P Rothwell, Acc. Chem. Res., **21**, 153 (1988); (f) R.H. Crabtree, Chem. Rev., **85**, 245 (1985); (g) J.A. Martinho Simoes and J. Beauchamp, Chem. Rev., **90**, 629 (1990)
40. (a) B.J. Burger, M.E. Thompson, W.D. Cotter and J.E. Bercaw, J. Am. Chem. Soc., **112**, 1566 (1990); (b) M.E. Thompson, S.M. Buxter, A.R. Bulls, B. Burger, M.C. Nolan, B.D. Santarsiero, W.P. Schaefer and J. Bercaw, J. Am. Chem. Soc., **109**, 203 (1987); (c) G. Jeske, H. Lauke, H. Mauermann, H. Schumann and T.J. Marks, J. Am. Chem. Soc., **107**, 8111 (1985); (d) J.W. Bruno, G.M. Smith, T.J. Marks, C.K. Fair, A.T. Shultz and J.M. Williams, J. Am. Chem. Soc., **108**, 40 (1986); (e) J.A. Davis, P.L. Watson, J.F. Liebman and A. Greenberg, Eds., *S*elective Hydrocarbon Activation, VCH Publishers, New York (1990); (f) C. McDade and J.E. Bercaw,

J. Organomet. Chem., **279**, 281 (1985); (g) A.R. Bulls, J.E. Bercaw, J.M. Marinquez and M.E. Thompson, Polyhedron, **7**, 5750 (1988)

41. (a) C.S. Christ Jr, J.R. Eyler and D.E. Richardson, J. Am. Chem. Soc., **112**, 596 (1990); (b) C.S. Christ Jr, J.R. Eyler and D.E. Richardson, J. Am. Chem. Soc., **110**, 4038 (1988)
42. (a) K. Kitaura, S. Obara and K. Morokuma, J. Am. Chem. Soc., **103**, 2891 (1981); (b) S. Obara, K. Kitaura and K. Morokuma, J. Am. Chem. Soc., **106**, 7482 (1984); (c) J.O Noell and P.J. Hay, J. Am. Chem. Soc., **104**, 4578 (1982); (d) J.J. Low and W.A. Goddard, J. Am. Chem. Soc., **106**, 6928 (1984); (e) M.R.A. Blomberg and P.E.M. Siegbahn, J. Chem. Phys., **78**, 986 (1983); (f) M.R.A. Blomberg and P.E.M. Siegbahn, J. Chem. Phys., **78**, 5682 (1983); (g) U.B. Brandemark, M.R.A. Blomberg, L.G.M. Pettersson and P.E.M. Siegbahn, Phys. Chem., **88**, 4617 (1984); (h) J.J. Low and W.A. Goddard, J. Am. Chem. Soc., **106**, 8321 (1984); (i) J.J. Low and W.A. Goddard, Organometallics, **5**, 609 (1986); (k) J.J. Low and W.A. Goddard, J. Am. Chem. Soc., **108**, 6115 (1986); (l) J.-Y. Saillard and R. Hoffmann, J. Am. Chem. Soc., **106**, 2006 (1984)
43. (a) P. Hofmann and M. Padmanabhan, Organometallics, **2**, 1273 (1983); (b) A. Dedieu, *T*opics in Physical Organometallic Chemistry; Gielen, M. F., Ed., Vol 1, Freund Publishing House, London, 1 (1989); (c) T. Ziegler, L. Fan, V. Tschinke and A. Becke, J. Am. Chem. Soc., **111**, 2018 (1989); (d) T. Ziegler, L. Fan, V. Tschinke and A. Becke, J. Am. Chem. Soc., **109**, 1351 (1987); (e) T. Ziegler, C. Wendan, E.J Baerends and W. Ravenek, Inorg. Chem., **27**, 3458 (1988); (f) T. Ziegler, V. Tschinke, L. Versluis and E.J. Baerends, Polyhedron, **7**, 1625 (1988)
44. (a) M.L. Steigerwald and W.A. Goddard, J. Am. Chem. Soc., **106**, 308 (1984); (b) E. Folga, T. Ziegler and L. Fan, New. J. Chem., **15**, 741 (1991); (c) E. Folga and T. Ziegler, Can. J. Chem., **70**, 333 (1992)
45. L. Versluis and T. Ziegler, Organometallics, **9**, 2985 (1990)
46. P.L. Watson, *S*elective Hydrocarbon Activation, J.A. Davis, P.L. Watson, J.F. Liebman and A. Greenberg, Eds., VCH Publishers, New York (1990)
47. T. Ziegler and E. Folga, work in progress
48. (a) J. Feldman, W.M. Davis and R.R. Schrock, Organometallics, **8**, 2266 (1989); (b) R.R. Schrock, R.T. DePue, J. Feldman, K.B. Yap, D.C. Yang, W.M. Davis, L. Park, M. DiMare, M. Schofield, J. Anhaus, E. Walborsky, E. Evitt, C. Kruger and P. Betz, Organometallics, **9**, 2262 (1990); (c) J. Feldman, W.M. Davis, J.K. Thomas and R.R. Schrock, Organometallics, **9**, 2535 (1990); (d) R.R. Schrock, Acc. Chem. Res., **23**, 158 (1990); (e) R.R. Schrock, J.S. Murdzek, G.C. Bazan, J. Robbins, M. DiMare and M. O'Regan, J. Am. Chem. Soc., **112**, 3875 (1990); f) R.R. Schrock, R.T. DePue, J. Feldman, C.J. Schaverien, J.C. Devan and A.H. Liu, J. Am. Chem. Soc., **110**, 1423 (1988)
49. O. Eisenstein, R. Hoffmann and A.R. Rossi, J. Am. Chem. Soc., **103**, 5582 (1981)
50. (a) A.K. Rappé and T.H. Upton, Organometallics, **3**, 1440 (1984); (b) T.H. Upton and A.K. Rappé, J. Am. Chem. Soc., **107**, 1206 (1985); (c) A.K. Rappé and W.A. Goddard III, J. Am. Chem. Soc., **102**, 5115 (1980); (d) A.K. Rappé and W.A. Goddard III, J. Am. Chem. Soc., **104**, 3287 (1982)
51. T.R. Cundari and M.S. Gordon, Organometallics, **11**, 55 (1992)
52. K.P.C. Vollhardt, *Organic Chemistry*, H. F. Freeman and Company, New York, 900 (1987)
53. J.H. Wengrovius, J. Sancho and R.R. Schrock, J. Am. Chem. Soc., **103**, 3931 (1981)
54. L.G. McCullough, R.R. Schrock, J.C. Dewan and J.S. Murdzek, J. Am. Chem. Soc., **107**, 5987 (1985)
55. M.R. Churchill, J.W. Ziller, J.H. Freudenberger and R.R. Schrock, Organometallics, **3**, 1554 (1984)

56. J.H. Freudenberger, R.R. Schrock, M.R. Churchill, A.L. Rheingold and J.W. Ziller, Organometallics, **3**, 1563 (1984)
57. (a) M.L. Listmann and R.R. Schrock, Organometallics, **4**, 74 (1985); (b) G.J. Leigh, M.T. Rahman and D.R.M. Walton, J. Chem. Soc., Chem. Commun., 541 (1982)
58. R.R. Schrock, S.F. Pedersen, M.R. Churchil and J.W. Ziller, Organometallics, **3**, 1574 (1984)
59. R.R. Schrock, D.N. Clark, J. Sancho, J.H. Wengrovious, S.M. Rocklage and S.F. Pedersen, Organometallics, **1**, 1654 (1982)
60. (a) M.R. Churchill, J.C. Fettinger, L.G. McCullough amd R.R. Schrock, J. Am. Chem. Soc., **106**, 3356 (1984); (b) M.R. Churchill and J.C. Fettinger, J. Organomet. Chem., **290**, 375 (1985)
61. R.R. Schrock, J.S. Murdzek, J.H. Freudenberger, M.R. Churchill and J.W. Ziller, Organometallics, **5**, 25 (1986)
62. R.R. Schrock, Acc. Chem. Res., **19**, 342 (1986)
63. (a) E.V. Anslyn, M.J. Brusich and W.A. Goddard III, Organometallics, **7**, 98 (1988); (b) B.E. Bursten, J. Am. Chem. Soc., **105**, 121 (1983); (c) E.D Jemmis and R. Hoffmann, J. Am. Chem. Soc., **102**, 2570 (1980)
64. (a) G.A. McDermott, A.M. Dorries and A. Mayr, Organometallics, **6**, 925 (1987); (b) D.N. Clark and R.R. Schrock, J. Am. Chem. Soc., **100**, 6774 (1978); (c) A. Mayr, M.A. Kjelsberg, K.S. Lee, M.F. Asaro and T. Hsieh, Organometallics, **6**, 2610 (1987); (d) A. Mayr, M.A. Kjelsberg, K.S. Lee and D. Van Engen, Organometallics, **6**, 432 (1987)
65. M.G.B. Drew, B.J. Brisdon and A. Day, J. Chem. Soc., Dalton Trans., 1310 (1981)
66. (a) L.G. McCullough and R.R. Schrock, J. Am. Chem. Soc., **106**, 4067 (1984); (b) D. Villemin and P. Cachiot, Tetrahedron Lett., **15**, 93 (1982)
67. T.A. Albright, J.K. Burdett and M.H. Whangbo, *O*rbital Interactions in Chemistry, John Wiley & Sons, New York (1984)
68. (a) M.R. Churchill, J.W. Ziller, L. McCullough, S.F. Pedersen and R.R. Schrock, Organometallics, **2**, 1046 (1983); (b) M.R. Churchill and J.W. Ziller, J. Organomet. Chem., **279**, 403 (1985)

WACKER REACTIONS

Introduction

Among the homogeneous catalytic processes used industrially, the Wacker process, which produces acetaldehydes by the oxidation of ethylene using Palladium(II) chloride and Copper(II) chloride as catalyst, has been the subject of many mechanistic studies since its discovery in 1958 [1]. The overall picture of the reaction mechanism may be summarized, at least for the reaction occurring in water, in the set of equations *1–3* shown below:

(1) $C_2H_4 + H_2O + PdCl_2 \rightarrow CH_3CHO + Pd + 2\ HCl$

(2) $Pd + 2\ CuCl_2 \rightarrow PdCl_2 + 2\ CuCl$

(3) $2\ CuCl + 2\ HCl + 1/2\ O_2 \rightarrow 2\ CuCl_2 + H_2O$

$C_2H_4 + 1/2\ O_2 \rightarrow CH_3CHO$

Yet most of the mechanistic work carried out so far on this process has been devoted to the first equation, the formation of acetaldehyde coupled with the reduction of Pd(II) to Pd(0). Little is known on the other hand [2] about the actual mechanism of the two subsequent steps which deal with the reoxidation of the Pd metal by Cu(II) chloro complexes, thus making the whole process catalytic. For the acetaldehyde formation (equation *1*), the following scheme has been proposed on the basis of kinetic and deuterium labeling studies and isotope effects [3].

(4) 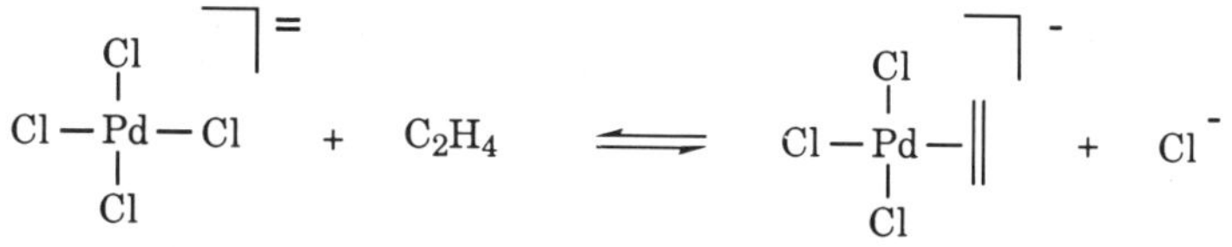 $[PdCl_4]^{2-} + C_2H_4 \rightleftharpoons [PdCl_3(C_2H_4)]^- + Cl^-$

(5) 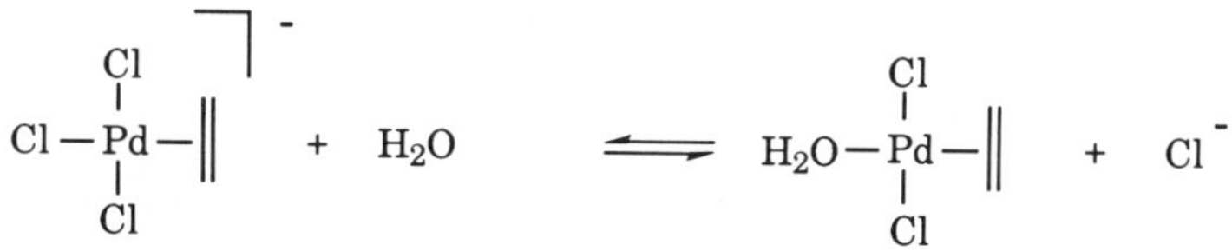 $[PdCl_3(C_2H_4)]^- + H_2O \rightleftharpoons PdCl_2(H_2O)(C_2H_4) + Cl^-$

P.W.N.M. van Leeuwen et al. (eds.), Theoretical aspects of homogeneous catalysis, 167–195.

$$(6)\quad [PdCl_2(H_2O)(C_2H_4)] + H_2O \rightleftharpoons [PdCl_2(H_2O)(CH_2CH_2OH)]^- + H^+$$

$$(7)\quad [PdCl_2(H_2O)(CH_2CH_2OH)]^- \xrightarrow{\text{slow}} CH_3CHO + Pd(0) + 2\,Cl^- + H^+$$

In this scheme the slow step (equation *7*) is the reductive elimination of acetaldehyde. It may be decomposed as follows, the chloride dissociation step *8* being the rate determining step.

The proposal for the series of the β-elimination/insertion sequences (equations *9–11*) leading finally to the acetaldehyde product is based on the absence of any deuterium incorporation when the reaction is carried out in D_2O.

A comprehensive theoretical study of this process should therefore address all the steps listed above, or at least the most critical ones, e.g. the ethylene coordination step, the nucleophilic addition on the coordinated ethylene, and

$$(8)\quad [PdCl_2(H_2O)(CH_2CH_2OH)]^- \longrightarrow PdCl(H_2O)(CH_2CH_2OH) + Cl^-$$

$$(9)\quad PdCl(H_2O)(CH_2CH_2OH) \longrightarrow PdHCl(H_2O)(CH_2{=}CHOH)$$

$$(10)\quad PdHCl(H_2O)(CH_2{=}CHOH) \longrightarrow PdCl(H_2O)(CH(OH)CH_3)$$

$$(11)\quad (Cl)(H_2O)Pd{-}CH(OH)(CH_3) \longrightarrow (Cl)(H_2O)Pd{-}H + CH_3CHO$$

$$(12)\quad (Cl)(H_2O)Pd{-}H \longrightarrow HCl + H_2O + Pd(0)$$

the sequence of β-elimination/insertion steps. As yet, there is a long way to go. In fact the nucleophilic addition step *6* is the only one which has been studied thoroughly within the context of the Wacker process. One should notice however that the various theoretical studies that have been devoted to substitution processes in d^8 square planar complexes may be used to analyze the chlorine/ethylene and chlorine/water substitution steps. Such an analysis might be useful considering the fact that the olefinic complex has never been isolated. The purpose of the present chapter is to review the theoretical work performed on this process and to delineate what remains to be done in this field.

The nucleophilic addition step

The first theoretical study of the nucleophilic addition to a coordinated olefin may be traced to Eisenstein and Hoffmann [4], who carried out a detailed molecular orbital analysis within the framework of the Extended Hückel theory. The central argument of their analysis is that the activation of the olefin towards the nucleophilic addition does *not* result from the binding to the transition metal atom but rather from the slipping of the coordinated olefin from an η^2 to an η^1 coordination mode. It has been stressed however [5, 6] that the extended Hückel method might bias the argument. Self-consistent field approaches [6–8] performed later have yielded more quantitative result which in combination with the qualitative ones provide us now with a quite deep understanding of the nucleophilic addition on the coordinated ethylene in square planar Pd(II) complexes.

Let us begin our analysis of this step by looking at the coordination of the ethylene to the Pd(II) transition metal complex fragment. If one considers

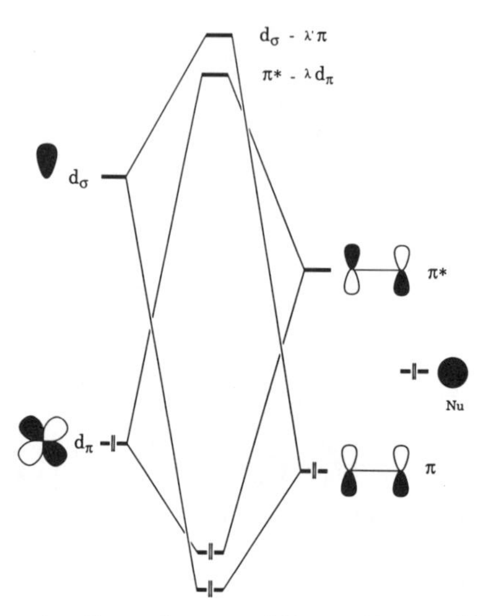

Fig. 1. Schematic interaction diagram between a ML_n transition metal fragment and ethylene. The σ orbital of the incoming nucleophile is arbitrarily placed between the π and π^* orbitals of C_2H_4.

the molecular orbital interactions only (as given for instance by the Extended Hückel calculations of Eisenstein and Hoffmann [4]), one ends up with a *deactivation* rather than with an activation of C_2H_4. This is a quite surprising feature (at least at first sight), which is best understood from the interaction diagram drawn for a $ML_n(C_2H_4)$ system, see Figure *1*. The empty π^* orbital of ethylene, which is the acceptor orbital for the incoming nucleophile, is destabilized through its interaction with a filled dπ orbital (which is usually present) of the ML_n fragment. The resulting $\pi^* - \lambda d_\pi$, although being mostly localized on the ethylene ligand is no longer pure. Thus from both the decrease in the overlap with the lone pair of the incoming nucleophile (either H_2O or OH^-) and the increase in the energy gap between this lone pair and $\pi^* - \lambda d_\pi$ one expects a decrease in the reactivity towards nucleophilic addition. Eisenstein and Hoffmann noticed however that in the case of d^8 $PdL_3(C_2H_4)$ systems (as the ones involved in the Wacker process) the deactivation is weak since the destabilization of the $\pi^* - \lambda d_\pi$ orbital is somewhat offset by an additional stabilization of the p orbital of the metal. As shown in **1**, this p orbital of relatively low energy,

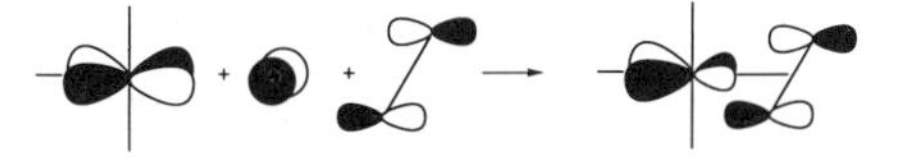

1

mixes in phase into the $\pi^* - \lambda d_\pi$ orbital. As a result this orbital has nearly the same energy as the π^* orbital in the free olefin.

The reader, familiar with the Dewar-Chatt-Duncanson bonding model of olefin metal complexes [9] would argue that the forward donation from the olefin π orbital into an empty d_σ (or p_σ) orbital of the metal, see **2** and the Figure *1*, has not been

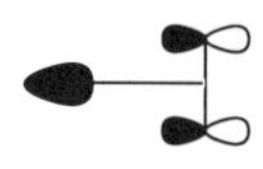

2

used in the above rationalization. This interaction, which depopulates the olefin π orbital, leads to a buildup of positive charge on the olefin, and hence can induce a lowering of all orbital energies, in particular a lowering of the π^* orbital energy. As a result the electrophilicity of the olefin should be increased. But one faces here one of the shortcomings of the Extended Hückel method, namely its inability to account for these features. Being an electron density independent method it is inadequate to model electrostatic type interactions or as here, to reproduce shifts of the orbital energies upon charge modification. In contrast Self-Consistent Field methods do account for such inductive effects which might well overcome the deactivation brought by the π back donation effects, especially if the latter are not very strong, as for instance in Pd(II) complexes [10]. Indeed INDO calculations (which are also semi empirical calculations but of self-consistent type) carried out for the nucleophilic addition of H^- on $[CpFe(CO)_2(C_2H_4)]^+$ do show the increase of the positive charge of the carbon atoms and the lowering of the C_2H_4 orbital energies, pointing to the importance of these inductive effects [6]. On the other hand and as argued by Eisentein and Hoffmann themselves, one should be aware of the fact that all calculations which have been carried out so far, including the SCF ones, correspond to hypothetical gas phase molecules and might therefore exaggerate the changes in charges and the corresponding inductive effects to be found in solution. Therefore the mere coordination of C_2H_4 may be not sufficient, at least in solution, to promote the nucleophilic addition.

Eisenstein and Hoffmann found instead that a slipping of C_2H_4 from the η^2 to the η^1 coordination mode is a key component of the activation towards the nucleophilic addition. Their argument may be summarized as follows: the lowering of the symmetry (or pseudosymmetry) of the system on going from η^2 to η^1 allows a mixing of the $d_\sigma - \lambda'\pi$ orbital into the $\pi^* - \lambda d_\pi$ orbital. Since the $d_\sigma - \lambda'\pi$ orbital lies generally above the $\pi^* - \lambda d_\pi$ orbital, see Figure *1*, this mixing polarizes the $\pi^* - \lambda d_\pi$ orbital in the way shown in **3**. The extent of the admixture is governed by the energy

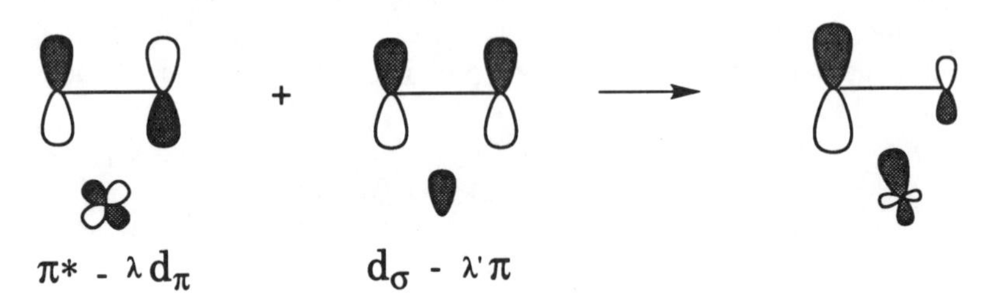

gap between the two orbitals: the lower the $d_\sigma - \lambda'\pi$ orbital is, the greater its mixing into $\pi^* - \lambda d_\pi$ is and thus the greater the activation of the ethylene is. One therefore expects that electron withdrawing substituents, which do not

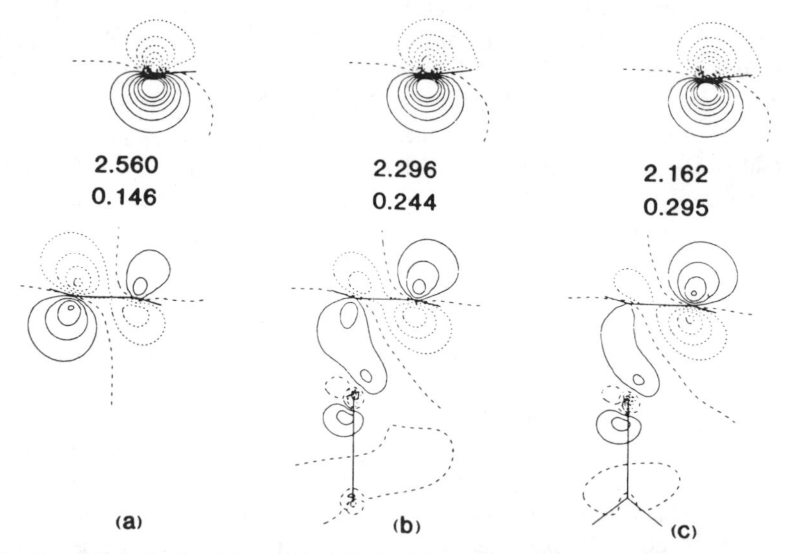

Fig. 2. Principal pairs of interacting orbitals of the systems (a) [OH . . . C_2H_4], (b) [OH . . . $PdCl_3(C_2H_4)$] and (c) [OH . . . $Pd(H_2O)Cl_2(C_2H_4)$]. The numbers indicate the electron population (upper) and the overlap population (lower) for the pairs of orbitals (reproduced with permission from ref. 7).

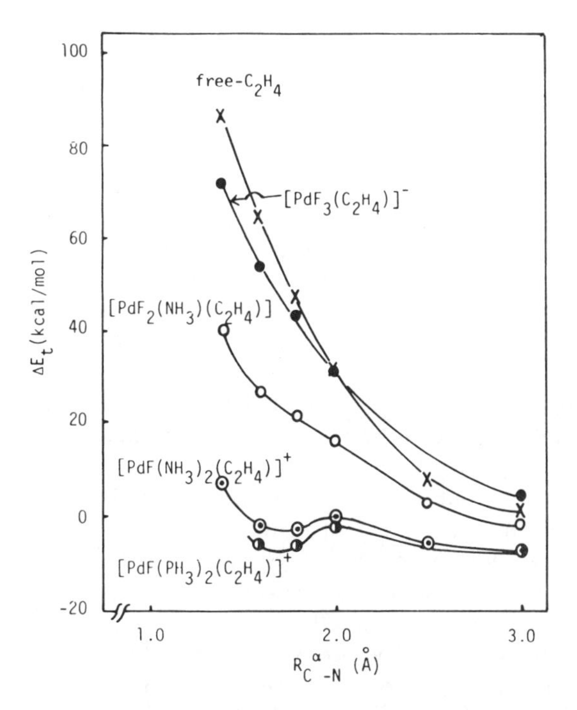

Fig. 3. Energy profile for the NH_3 nucleophilic attack on C_2H_4, $PdF_3(C_2H_4)$, $PdF_2(NH_3)(C_2H_4)$, $PdF(NH_3)_2(C_2H_4)^+$, and $PdF(PH_3)_2(C_2H_4)^+$ (reproduced with permission of ref. 8).

push the d_σ orbital too high in energy, will activate the nucleophilic addition, whereas good σ donor ligands will be less efficient, especially when lying *trans* to the olefin. At this stage of the discussion, it is perhaps worthwhile to mention that these ideas have got some experimental support: the X-ray crystal structure of dichloro(*endo*-dicyclopentadiene)palladium(II) shows one unsymmetrically bound double bond, with one carbon atom closer to the metal and one farther away being attacked by nucleophiles [11].

The EH results have been put on a more quantitative basis by Fujimoto and Yamasaki [7] in their coupled fragment molecular orbital method [12]. This method which basically consists of localizing (within the framework of the SCF theory) the orbitals of the interacting fragments in pairs of interacting and non interacting orbitals, not only connects *ab initio* MO-SCF calculations with the orbital interaction concept, but also quantifies the interaction by relying on the electron population and on the overlap population of the interacting orbitals: A value of 2 for the electron population will correspond to the optimum value to form an electron pair. The greater the deviation from 2 towards larger values, the greater the closed shell repulsion is. The overlap population of the interacting orbitals also provides a useful index of the amount of interaction. The results for the $[OH^- \ldots C_2H_4]$, $[OH^- \ldots PdCl_3(C_2H_4)]^-$ and $[OH^- \ldots Pd(H_2O)Cl_2(C_2H_4)]$ systems, see the Figure *2*, point clearly to the activation of the ethylene when coordinated to a Pd(II)L_3 fragment and to a greater activation in *trans*-$PdCl_2(H_2O)(\eta^1\text{-}C_2H_4)$ as compared to $[PdCl_3(\eta^1\text{-}C_2H_4)]^-$: The electron population of the interacting orbital made of the bonding combination between the π^* orbital of ethylene and the lone pair of OH^- decreases from 2.560 to 2.162 in the series, whereas the corresponding overlap population increases from 0.146 to 0.295. One sees also clearly in Figure *2* an increasing polarization of the π^* orbital on the C_β atom, whose population increases from 15% to 48% and 65%. Here too, and as in the Hoffmann and Eisenstein analysis, the difference in the catalytic activities between $PdCl_3^-$ and $PdCl_2(H_2O)$ is explained by a greater intermixing of π and π^* due to a greater involvement of the vacant $4d_{x^2-y^2}$ orbital (*i.e.* the d_σ type orbital).

Thus the η^1 coordination mode of C_2H_4 appears to be indeed activating towards the nucleophilic addition. The next step is to assess the origin of the driving force for the slipping of the ethylene from η^2 to η^1. This question has been addressed by Sakaki and coworkers [8] in a study – based on *ab initio* SCF calculations – which up to now is the most comprehensive one on this subject. Although these authors have not explicitly treated the nucleophilic addition of OH^- or H_2O (which are the actual nucleophiles involved in the

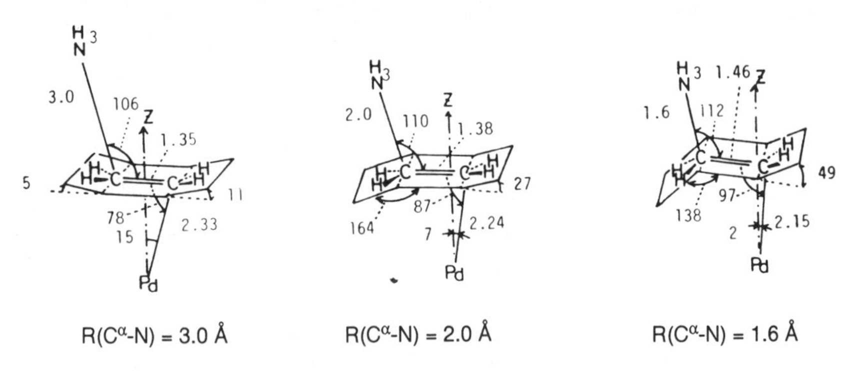

Fig. 4. Evolution of the geometry along the reaction path of the $NH_3 + PdF_2(NH_3)(C_2H_4)$ reaction (bond lengths in Å and angles in degrees). The ligands are omitted for the sake of clarity (reproduced with permission of ref. 8).

Wacker process), but rather the addition of NH_3, their analysis encompasses a variety of palladium metal complexes ($(PdF_3(C_2H_4)^-$, **4**, $PdF_2(NH_3)(C_2H_4)$, **5**, $PdF(NH_3)_2(C_2H_4)^+$, **6**, and $PdF(PH_3)_2(C_2H_4)^+$, **7**) and provides therefore an answer to many aspects of this reaction.

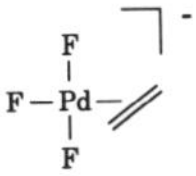

4 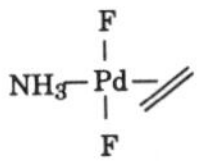 **5**

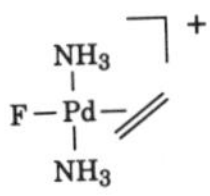

6 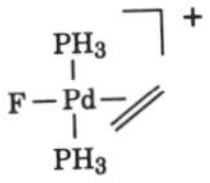 **7**

In all cases the reaction path was determined using a non-gradient approach. Three features are worth mentioning. First, the free C_2H_4, the neutral and the anionic ethylene complexes yield repulsive energy curves for the approach of the NH_3 nucleophile, see the Figure *3*. On the other hand the cationic systems $PdF(NH_3)_2(C_2H_4)^+$, **6**, and $PdF(PH_3)_2(C_2H_4)^+$, **7** give rise to a reaction channel with a moderate barrier (from 6 to 8 kcal/mol), in agreement with the experimental features known about the aminopalladation of olefins. Second, a quite important charge redistribution takes place during the attack: electron density is transferred, as schematically shown in **8**, from the β carbon atom of the coordinated ethylene to

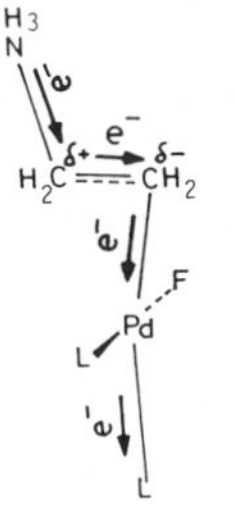

8

(With permission of [8])

palladium and to the ligand *trans* to it. Third, as shown in the Figure *4*, the geometry changes occurring during the NH_3 approach correspond to a change of the coordination mode of ethylene from η^2 to η^1. An energy decomposition analysis (in which the electrostatic, the exchange, the forward charge trans-

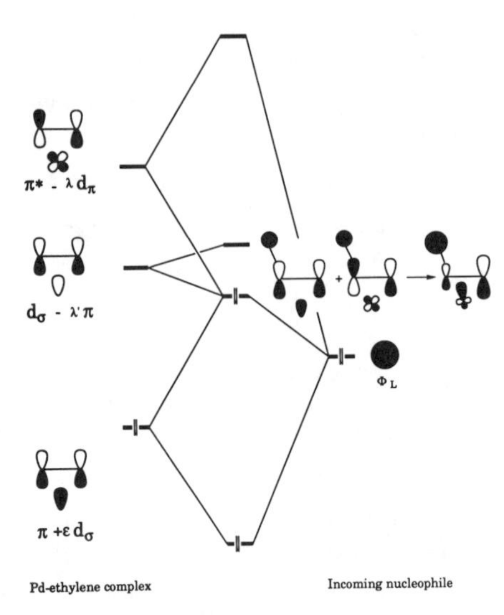

Fig. 5. Schematic orbital interaction diagram between the Pd-ethylene complex and the incoming nucleophile.

fer and the backward charge transfer contributions to the interaction energy between the nucleophile and the substrate can be singled out) was also performed. It indicates that the reaction is simultaneously frontier and charge controlled: the smaller the energy difference between the π^* orbital of C_2H_4 and the lone pair of the nucleophile, the easier the reaction is. The π^* orbital energy is in turn governed by the overall charge of the system and the nature of the ancillary ligands, in particular by their propensity to accept electron density. Thus by pushing up the $\pi^*C_2H_4$ orbital anionic ligands disfavor the metal assisted nucleophilic attack. On the other hand neutral and soft ligands keep $\pi^*C_2H_4$ quite low and can themselves accept electron density during the attack, thereby favoring the nucleophilic attack. One finds again the results obtained by Eisenstein and Hoffmann and by Fujimoto and Yamasaki. It is therefore clear from all these studies that the substitution of a Cl ligand by H_2O (equation *5*) is a necessary and important step for the activation of the coordinated ethylene.

One might worry about the fact that the neutral systems **5** and **7** give rise to a *repulsive* potential energy curve, because the neutral system is the one, *in the catalytic process*, which is believed to undergo the nucleophilic attack. In study of Sakaki however the nucleophile is NH_3 instead of OH^- and the greater basicity of OH^- [13, 14] should lead to an attractive potential energy curve.

In the two previous studies the rationalization of the reactivity features was based on frontier orbital theory, thus offering a quite static view of the process. In contrast the study of Sakaki *et al.* analyzes the orbital mixing during the nucleophile approach. One therefore gets a more dynamic picture. As before, a mixing of the π and π^* orbitals (or more precisely of the $\pi^* - \lambda d_\pi$ and $\pi + \varepsilon d_\sigma$ orbitals) of the coordinated C_2H_4 is involved. But now the mixing is induced by the lone pair Φ_L of the incoming nucleophile, see Figure *5*. A mixing of the $d_s - \lambda'\pi$ is of course also found but it does not dominate the orbital interaction pattern. The net result of the π/π^* mixing is an enlargement of the p_π lobe on the C_β atom at the expense of the p_π lobe on the C_α atom. One therefore expects a strengthening of the bonding interaction between Pd and C_β and a weakening of the bonding between Pd and C_α. The driving force for the slipping is therefore induced by the incoming nucleophile. This mixing and hence the driving force is maximum for cationic systems (owing to the smallest energy difference between Φ_L and $\pi^* - \lambda d_\pi$. When going to neutral and anionic complexes, the energy difference between Φ_L and $\pi^* - \lambda d_\pi$ increases (because the orbitals of the Pd complex are pushed up), the admixture of π^* into π is less important and accordingly the driving force for the slipping is reduced.

A final question dealing with the nucleophilic addition step and which has been the subject of many controversies is whether the addition to the coordinated ethylene needs the intermediacy of a complex with a coordinated OH (*cis* addition, see Scheme *1*), or is a direct nucleophilic addition (*trans* addition, Scheme *1*). This

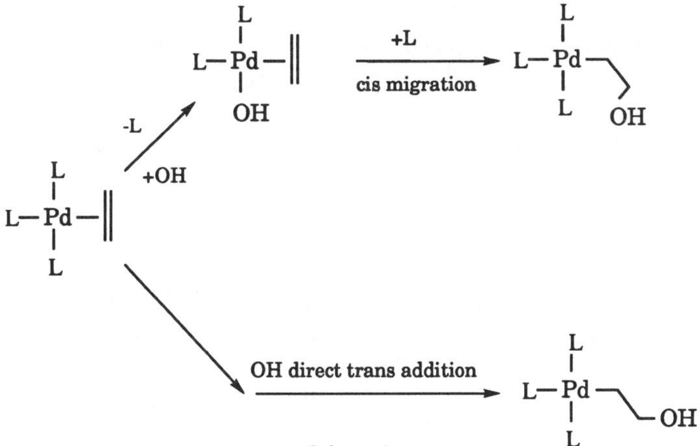

L
L—Pd—‖
L
-L
+OH
L
L—Pd—‖
OH
+L
cis migration
L
L—Pd
L
OH
OH direct trans addition
L
L—Pd
L
OH

Scheme 1.

question has now been experimentally settled in favor of the *trans* addition [3c] and a theoretical study based on *ab initio* SCF calculations and frontier orbital theory arguments has rationalized the experimental observation [15] by correlating the nucleophilic behaviour of various nucleophiles (Nu = H^-, CH_3^-, OH^- and F^-) to the orbital energy difference between the $\sigma_{Pd\text{-}Nu}$ bonding orbital and the π^* orbital in the *trans*-$Pd(Nu)_2(H_2O)(C_2H_4)$ square planar system. In the case of Nu = H^- and CH_3^- a rather small energy gap was found, much smaller (by about 7eV) than the one found for the harder nucleophiles OH^- and F^-. This criterion has been put forth to explain the ease of the *cis*-migration in the H^- and CH_3^- case. One should caution the reader however that this analysis overlooked the possible involvement of the π–C_2H_4 orbital which is known now to be non negligible in the transition state of both the *cis* migration [16] and the *trans* addition (*vide supra*).

Before closing this section, let us summarize the main conclusions obtained from the various theoretical studies described above:

- the activation of the olefin towards the nucleophilic addition is due to its slipping from the η^2 to the η^1 coordination mode;
- this slipping is induced by the OH^- incoming nucleophile;
- the substitution of a Cl^- ligand by H_2O (which is a soft donor ligand) is also necessary to obtain a neutral ethylene complex which is more prone to the nucleophilic attack.

Substitution reaction of square planar Pd(II) complexes

Central to our previous discussion is the occurrence of the $PdCl_2(H_2O)(C_2H_4)$ complex and the sequence of substitutions leading to it (equations *4* and *5*). Yet this complex has never been observed. Its geometry (*cis* or *trans*) is unknown. In their study Sakaki *et al.* [8] carried out a geometry optimization of the $Pd(C_2H_4)$ unit for the $PdF_2(NH_3)(C_2H_4)$ system. The optimized Pd-C bond length (2.39 Å) points to a rather weak Pd-(C_2H_4) bond, the computed binding energy being 18 kcal/mol [8, 17]. The C_2H_4 ligand is also hardly deformed from its equilibrium structure: the hydrogen atoms bend back only slightly (7°) and the C-C bond length amounts to 1.34 Å. Similar results had been obtained by Hay in a previous calculation of $[PdCl_3(C_2H_4)]^-$ [18]: binding energy = 12.3 kcal/mol, Pd-C = 2.35 Å, C-C = 1.35 Å. The planar form was found even more weakly bound, binding energy = 5.4 kcal/mol, Pd-C = 2.77 Å, C-C = 1.34 Å. The computed values found for the upright form are in rough agreement with the ones found in X-ray crystal structures of similar Pd(II) square planar complexes [11, 19] except for the computed Pd-C bond lengths which seem to be too long. The weak binding has been ascribed by Sakaki [8] and Hay [18] to a weak back donative $d_\pi \rightarrow \pi^*$ interaction.

Should one trust these results which are obtained from rather primitive calculations, carried out at the SCF level only and with a geometry optimization limited to successive parabolic interpolations for the most important parameters? This question is all the more pertinent since calculations performed on the related system $[PtCl_3(C_2H_4)]^-$, where a direct comparison with experiment [20] is available, apparently suffer from the same drawbacks: the computed Pt-C bond length is again too long by 0.10 Å: (2.23 Å instead of 2.13 Å in the experimental structure) in the calculation of Hay [18] and by even more, 0.17 Å, in a more recent calculation of Sakaki, despite the use of a SCF energy gradient method [21]. Quite long Pd-C bond lengths have also been found when using gradient optimization techniques in related systems, e.g. 2.50 Å in $PdH_2(PH_3)(C_2H_4)$ [16b] or 2.46 Å in $Pd(CH_3)Cl(NH_3)(C_2H_4)$ [22]. Thus the use of a limited geometry optimization is not the cause of these difficulties. On the other hand calculations including electron correlation – in order to get a better description of the π back donation effects – and relativistic corrections might well yield larger binding energies and shorter Pd-C bond lengths: For the $[PtCl_3(C_2H_4)]^-$ system the inclusion of the correlation effects strengthen the binding of C_2H_4 by 9 to 15 kcal/mol depending on the method used [21], thus giving binding energies ranging around 30 kcal/mol. Calculations carried out in another context by Blomberg *et al.* on $Pd(C_2H_4)$ and $PdH_2(C_2H_4)$ have shown that relativistic corrections increase the binding of C_2H_4 by about 10 kcal/mol [23–25]. Clearly a thorough theoretical investigation of $[PdCl_3(C_2H_4)]^-$ would be worthwhile in order to get a *quantitative* picture of the bonding.

Having in mind the problems associated with the description of the binding of C_2H_4 to Pd(II), one may wonder about the fate of the substitution reactions *4* and *5*. To our knowledge, no theoretical study of these reactions has been carried out so far. Experimental kinetic data for related reactions are quite scarce. A very recent study of the ethylene exchange kinetics in $PdCl_3(C_2H_4)^-$ (but carried out in THF as a solvent) indicate that this exchange is first order with respect to free ethylene and is enthalpy controlled, the entropy term being relatively small and negative ($\Delta H^\ddagger = 12 \pm 1$ kcal/mol, $\Delta S^\ddagger = -5 \pm 4$ kcal/mol) [28]. Accordingly the authors of the study favor as reaction mechanism, by analogy with the $PtCl_3(C_2H_4)^-$ case, the formation of a trigonal-bipyramidal (TBP) activated complex $[PdCl_3(C_2H_4)_2]^-$ where the entering and leaving ethylene ligands and the trans chloride are in the equatorial plane. Thus a TBP arrangement might well also prevail for the activated complex $[PdCl_3(C_2H_4)(H_2O)]^-$ of equation *5*.

This feature seems to take place in most substitution reactions of square planar Pd(II) (or Pt(II)) complexes, as recently summarized by R.J. Cross [29]. Activation parameters experimentally found for such reactions generally point to associative reactions having 5-coordinate species as intermediates or transition states. We do not differentiate here between the two pathways available, either the k_1 solvation pathway or the k_2 ligand pathway, see Scheme *2*.

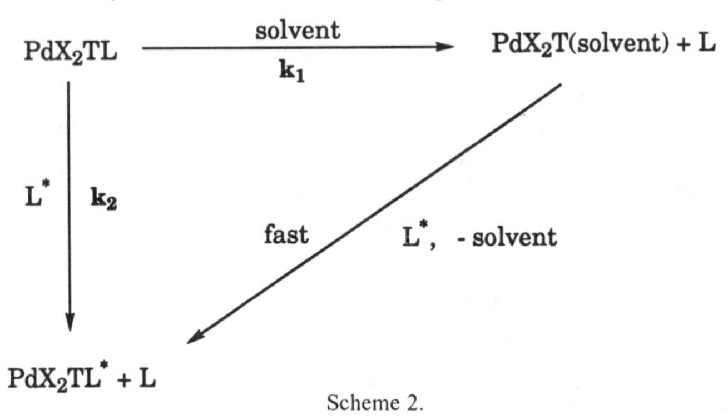

Scheme 2.

What is of interest to us at present and relying on the capabilities of the current theoretical methods, is the intimate nature of these associative reactions: in other words one would like to know the geometric and electronic structure of these 5-coordinate species, in order to rationalize or to predict an acceleration of the substitution and to assess whether they are intermediates or transition states. At this stage one should point out that it has been experimentally recognized [29, 30] that solvent interactions *at the metal site* are probably not energetically important. This does not imply that solvation itself is unimportant (one finds indeed changes in the k_2 term with solvents, like for instance in the $[PdCl_3(C_2H_4)]^-$ ethylene exchange reaction [28]), but it is probably of outer sphere nature (occurring on the coordinated ligands, on the entering and leaving ligands or through weak outer sphere complex formation) [29]. Consequently theoretical studies of such 5-coordinate species, which do not yet incorporate solvent effects, remain nevertheless quite useful.

A number of such studies have been carried out in the past, using empirical or semi empirical calculations. We have reviewed them in a previous article [31]. At that time the geometry of the 5-coordinate species could not be optimized with certainty and these studies relied instead very much on the previous experimental speculations. They gave nevertheless useful information about the *trans*-effect, i.e. the distinct preference for a site trans to one ligand rather than another. The early extended Hückel calculations of Zumdahl and Drago [32] carried out on the composite Cl^- + *trans*-$PtCl_2(L)(NH_3)$ systems (L = H^- or H_2O) used orbital overlap populations in the trigonal bipyramidal activated complex as an index: thus on going from H_2O (a poor σ donor) to H^- (a strong σ donor) the interaction **9** between the σ_L orbital of the trans ligand (either H_2O or H^-) with the Pt 6p orbital increased. This was correlated to a stronger stabilization of the 5-coordinate species in the H^- case.

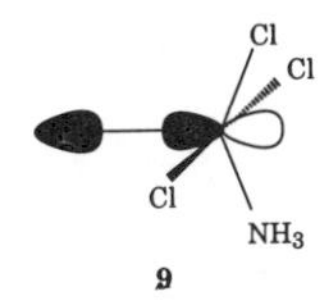

Similar conclusions had also been obtained in a CNDO study carried out later by Amstrong *et al.* [33], although these calculations favored the interaction of the σ_L orbital with the $5d_{x^2-y^2}$ and the 6s orbital over the one with the 6p orbital. The same study also addressed the rationalization of the *trans* effect found for π acceptor ligands. An approximate reaction path was determined for the self exchange reaction $[PtCl_3(C_2H_4)]^- + Cl^- \rightarrow [PtCl_3(C_2H_4)]^- +$ Cl^-. Assuming a trigonal bipyramidal transition state, a reaction path such as the one shown in the Scheme *3*

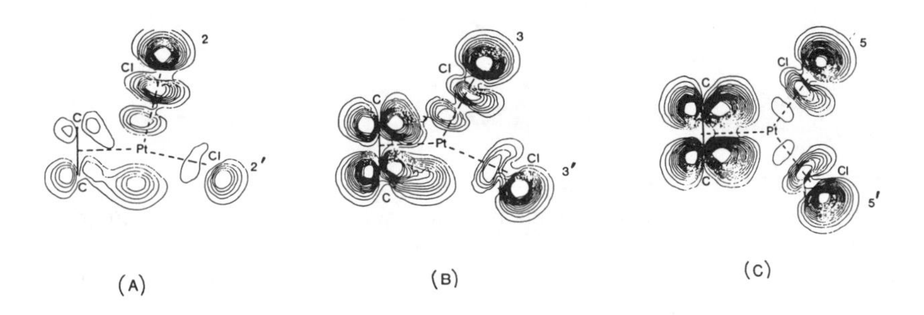

Fig. 6. Electron-density contours of the highest filled occupied orbital for three points on the reaction coordinate of the $[PtCl_3(C_2H_4)]^- + Cl^- \rightarrow [PtCl_3(C_2H_4)]^- + Cl^-$ self-exchange reaction (reproduced with permission of ref. 33).

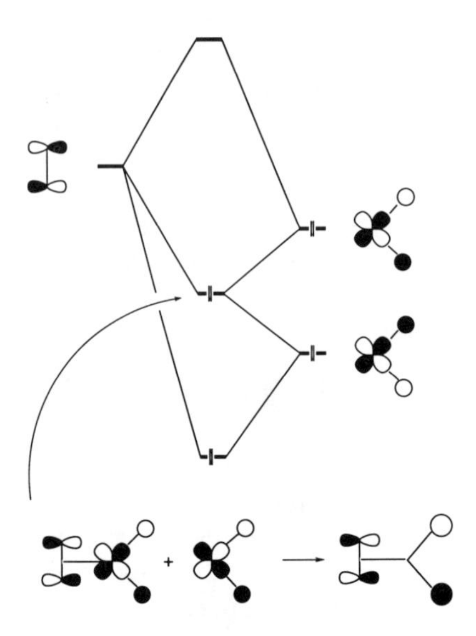

Fig. 7. Orbital interaction diagram between d_π, $\pi^*_{C_2H_4}$ and the out-of-phase combination of the two σ lone pairs of the entering and leaving ligands in the $[PtCl_3(C_2H_4)]^- \ldots Cl^-$ system.

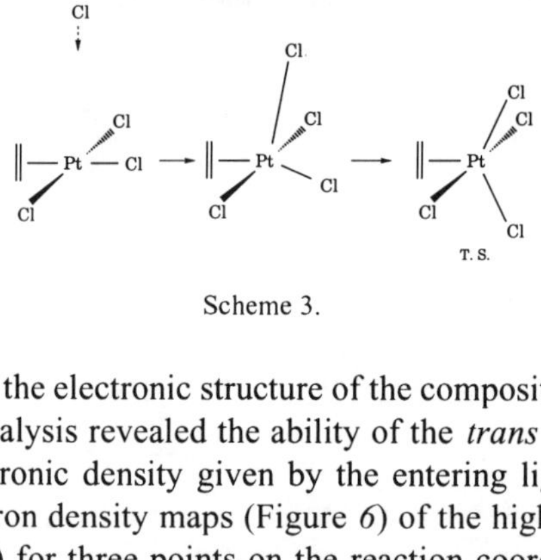

Scheme 3.

was considered and the electronic structure of the composite system analyzed throughout. This analysis revealed the ability of the *trans* ethylene ligand to delocalize the electronic density given by the entering ligand. This is best seen from the electron density maps (Figure *6*) of the highest filled molecular orbital (HOMO) for three points on the reaction coordinate, showing in particular the increase in the population of the $\pi^*_{C_2H_4}$ orbital. The delocalization is mediated by the d_π orbital which can interact with the orbitals of three ligands, provided that there is a trigonal or an approximate trigonal arrangement of the entering and leaving ligands with C_2H_4. The reader may

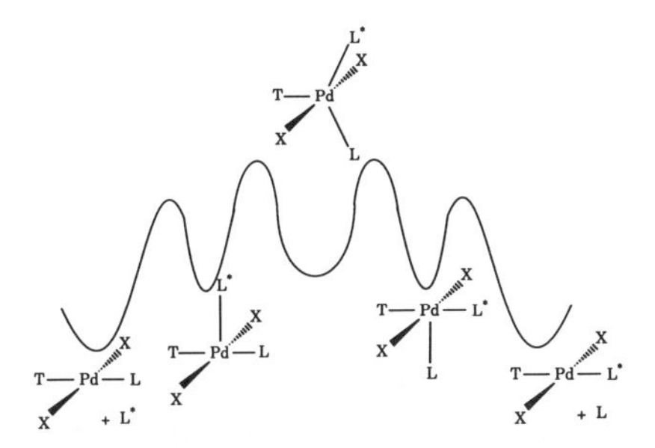

Fig. 8. A possible schematic energy profile for substitution reactions of square planar Pd(II) complexes.

have noticed that there is almost no electronic density associated with the d_π orbital in this HOMO. We think that this is the result of a three orbital mixing (between d_π, $\pi^*_{C_2H_4}$ and the out-of-phase combination of the two σ lone pairs of the entering and leaving ligands).

The corresponding orbital diagram of the Figure *7* shows how the electronic density of the two chloride ligands is delocalized on the $\pi^*_{C_2H_4}$ orbital *via* the d_π orbital and why a trigonal arrangement is necessary for the d_p orbital to act as a relay: in a square planar arrangement the s lone pairs of the entering and leaving ligands would not interact with the d_π orbital since they would be directed towards the nodal planes of this d_π orbital. At this point of the discussion one should recall that in their study of transition metal coordination Rossi and Hoffman showed that the preferential position of an ethylene ligand in a transition metal complex with a trigonal bipyramidal structure is in the equatorial plane [34].

The study of Amstrong *et al.* rationalizes in an elegant manner the trans effect of the π acceptor ligand. As far as the Wacker process is concerned it accounts therefore for the ease of the nucleophilic substitution step *5* where a chloride ligand is replaced by an H_2O ligand. One will expect (i) a stabilization of the corresponding transition state by the π accepting C_2H_4 ligand, and (ii) a *trans* disposition of H_2O and C_2H_4 in the intermediate, as we put on the Scheme *1*. We recall here that nothing is known experimentally about the stereochemistry of this intermediate which is the one that undergoes the nucleophilic addition.

The two previous studies made the assumption that the five coordinate species involved in the associative process is a transition state rather than an intermediate. Yet the fact that many five coordinate complexes of Pd(II) (and of Pt(II) as well) have been isolated and structurally characterized either as trigonal bipyramidal (TBP) or (and more frequently) as square pyramid (SP) [29] seems to indicate that the reaction coordinate for the ligand substitution may be as complicate as the one of Figure *8*, i.e. with energy minima corresponding to SP and TPB intermediates.

Some of these energy minima may of course disappear, depending on the actual systems. Interestingly, a double humped energy profile has been obtained by Burdett from calculations based on the angular overlap method [35] by assuming a symmetrical bending back of the leaving ligand and of its *trans* partner linked with the approach of the entering ligand. An effort was also made in this study to unravel the electronic factors governing the heights of the two barriers, relating the first barrier to the *trans* effect and the second barrier to the *trans* directing influence. According to these calculations the height of the entering barrier dominates most often the reaction rate. The interaction of the ligand orbitals with the (n + 1)s and (n + 1)p orbitals of the metal (5s and 5p in the case of palladium) is found to be important as in the study of Zumdahl and Drago [32]. The height of the entering barrier decreases (and hence the reaction rate should increase) with the decreasing σ strength and/or the π acceptor capability of the entering ligand, and with the σ strength of the *trans* ligand. On the other hand it does not depend on the π acceptor properties of the trans ligand. This last feature seems to be somewhat contradictory with the known experimental data [29], which show a marked depend-

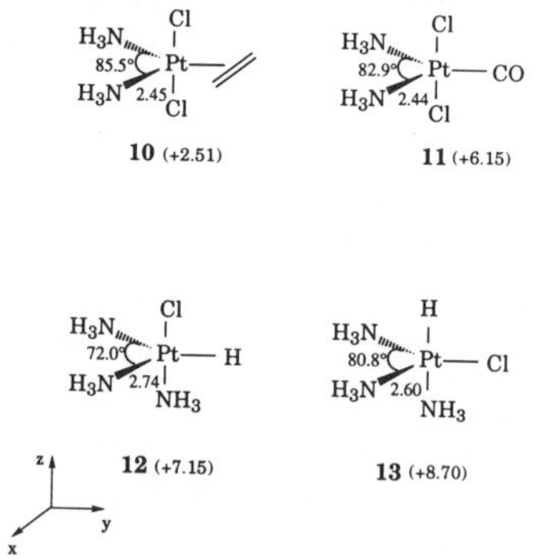

Fig. 9. Transition state geometries for the *trans*-Pt $(C_2H_4)Cl_2(NH_3)$ + NH_3, (**10**), *trans*-Pt $(CO)Cl_2(NH_3)$ + NH_3, (**11**), and *cis*-Pt $(H)Cl(NH_3)_2$ + NH_3 (**12, 13**) reactions.

ence of the *trans* effect upon the π acceptor properties of the *trans* ligand.

The height of the leaving barrier is, according to Burdett, a measure of the *trans* directing influence, and is found to increase with a decreasing σ donor strength and/or an increasing π acceptor strength of the trans ligand. This last point is therefore more in agreement with the CNDO calculations cited before [33], but we see here that the π acceptor capability of the trans ligand governs the trans directing influence rather that the *trans* kinetic effect.

These calculations, which were very useful at their time as an effort to rationalize the observed experimental features, suffered however from many limitations and drawbacks. Relying all on semi-empirical methods they were for instance parameter dependant, and this may explain why some results stress the interaction of the ligand orbitals with the nd metal orbitals while some others point to the importance of the (n + 1)s and (n + 1)p metal orbitals. Moreover in none of these studies had the reaction path been determined unequivocally, i.e. by using energy gradient minimization techniques based on rigorous *ab initio* calculations.

Very recently a study of this type has been carried out by Lin and Hall [36] on Pt and Rh complexes. We shall concentrate here on the Pt complexes which are more analogous to the palladium complexes involved in the Wacker process. The systems
trans-$Pt(C_2H_4)Cl_2(NH_3) + NH_3$
trans-$Pt(CO)Cl_2(NH_3) + NH_3$
cis- $Pt(H)Cl(NH_3)_2 + NH_3$
were chosen with identical entering and leaving NH_3 groups so that by applying the principle of microscopic reversibility, the geometry optimization of the five coordinate species could be done by looking for a global energy minimum with symmetry restriction, rather than by carrying out saddle point calculations. The corresponding geometries are shown in the Figure *9*, see **10–13**, together with the SCF energy barrier for the NH_3 self exchange. The reported values point to a kinetic trans effect in the order C_2H_4 > CO > H > Cl, in good agreement with the experimental order. Note also the very low barrier found for the NH_3 self exchange in *trans*-Pt $(C_2H_4)Cl_2(NH_3)$ which is most closely related to the systems involved in the Wacker process. The *trans* influence was also investigated through a comparison of the ligand dissociation energies, but unfortunately on the Rh complexes only. It is nevertheless worthwhile to point out that the dissociation energies were much higher than the energy barriers for the associative mechanism.

Lin and Hall also carried out a detailed analysis of the reaction path for two reactions, namely

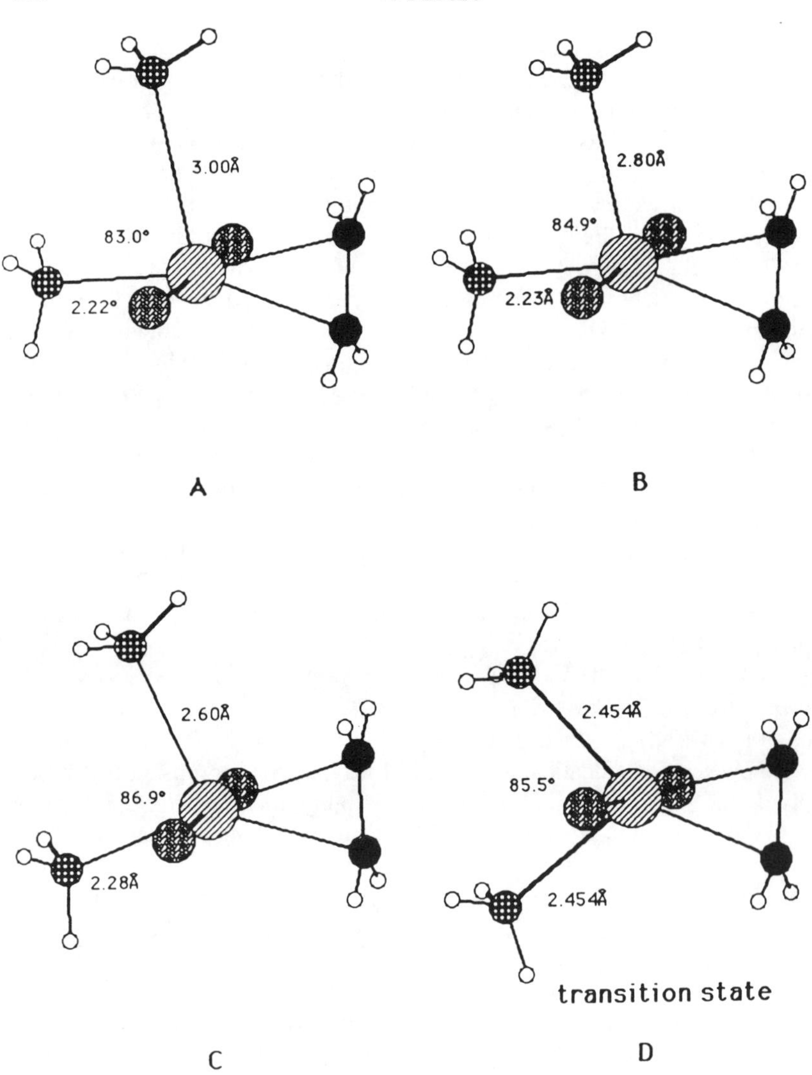

Fig. 10. Evolution of the geometry along the reaction path of the *trans*-$Pt(C_2H_4)Cl_2(NH_3)$ + NH_3reaction (reproduced with permission from ref. 36).

(13) *trans*-$Pt(C_2H_4)Cl_2(NH_3) + NH_3$ (transition state **10**)

(14) *cis*-$Pt(H)Cl(NH_3)_2 + NH_3$ (transition state **12**)

Figure *10* displays the evolution of the geometry along the reaction path for the reaction *13*. At the early stage of the reaction the attack occurs at the apex of the square plane, the electron pair of the entering ligand using the vacant p_z orbital. But when approaching further the palladium atom this electron pair experiences a repulsion from the occupied d orbitals. These occupied orbitals (d_{z^2}, **14**, d_{yz}, **15** and d_{xz}, **16**) have lobes which are directed mainly above and below the xy plane and give

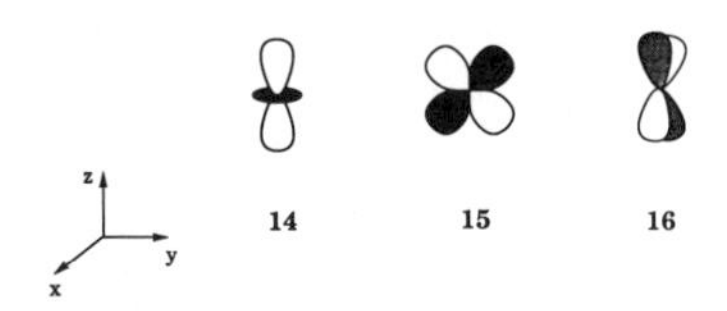

therefore rise to a concentration of electron charge in these regions and a depletion of electron charge in the xy plane. These electron charge concentration and depletion could be clearly seen in a plot of the Laplacian of the total valence electron density $-\nabla^2\rho$ [35]. Thus, in order to relieve part of this repulsion the entering ligand shifts away from the trans directing ligand and at the same time the leaving ligand is pushed down, away from the square plane. The system ends up in a pseudo trigonal bipyramidal transition state in which the angle between the entering and the leaving ligand is rather small (less than 90°, see **10–13**). This small angle is traced to the minimization of the repulsion between the electron pairs of the entering and leaving ligands and the metal occupied d orbitals. The minimization of this electron repulsion is in turn related to the properties of the trans directing ligand: a strong π accepting ligand (which occupy preferentially an equatorial position [34]) will stabilize (either directly by π back donation or indirectly because of the reduced charge of the metal) the two highest occupied orbital **17** and **18** which are derived from **14** and **15**.

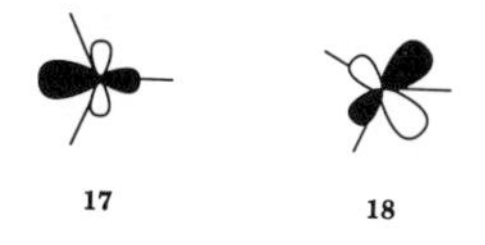

A strong σ donor by weakening the metal entering and metal leaving bonds decreases the electron repulsion. Thus both σ and π effects occurring in the transition state are involved in the trans effect.

A final point to be noticed is that Lin and Hall did not find any indication of a stable intermediate in these "gas-phase" calculations and they therefore speculated that the existence of an intermediate might be influenced by solvent effects or by different ligands. One such case, which they did not mention, would be the NH_3 self-exchange reaction $PtCl_2(NH_3)_2 + NH_3 \rightarrow PtCl_2(NH_3)_2 + NH_3$ where the three NH_3 ligands in the equatorial plane of the five coordinate species would be equivalent. If a transition state with two long Pt-N bonds, a corresponding small N-Pt-N angle and one short Pt-N bond would exist (similar to what is found in **10–13**) it would be interesting to know its position in the energy profile relative to a true trigonal bipyramidal $PtCl_2(NH_3)_3$ structure expected by symmetry.

The β-elimination/insertion steps

To our knowledge these steps – *8–12* – have not been studied *per se* theoretically. But theoretical *ab initio* studies of similar hydrogen transfer reactions have been performed on other systems. The corresponding results have been reviewed recently [16b] and are also the subject of the Chapter 3. They can be used to discuss some features of the steps *8–12*.

According to Bäckvall *et al.* [3c] the chlorine dissociation step *8* is the rate determining step of the overall process. As pointed out by these authors themselves this step, by providing a three coordinate Pd(II)hydroxyethyl complex makes the following β-H reductive elimination step *9* easier. This is because the elimination step will then end on with a four-coordinate rather than with a five-coordinate intermediate. It has been shown theoretically [37] and experimentally [38] that β-elimination usually requires coordinative unsaturation at the metal center. Moreover on the basis of the calculations which have been carried out so far, one expects that the activation barrier for the β-H elimination will be higher for four-coordinate alkyl systems than for three-coordinate alkyl systems. It amounts to about 34 kcal/mol in $Rh(PH_3)(CO)_2(C_2H_5)$ [16a] and to 11 kcal/mol only in $Pd(H)(PH_3)(C_2H_5)$ [39]. One has to be aware however that the β-elimination picture might be complicated by a rearrangement of the d^8 three-coordinate systems, between the Y shape and the T shape geometries [37, 40]. Also for a system such as $PdCl(H_2O)(CH_2CH_2OH)$, the presence of OH in the β-hydroxyalkyl ligand will influence the rate of the β-H elimination process. Fluorine substituents increase for instance the energy

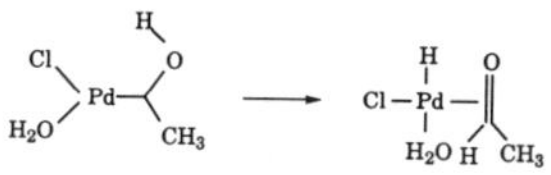

19

barrier of the β-elimination in $Pd(H)(PH_3)(CH_2CHF_2)$ with respect to the one in $Pd(H)(PH_3)(CH_2CH_3)$ by decreasing the electron donating ability of the $\sigma_{C\text{-}H}$ orbital [39]. Similarly the insertion reaction of the step *10* should be influenced by the OH group of the vinyl alcohol.

The first stage of the β-H elimination from the α-hydroxyethyl intermediate $PdCl(H_2O)(CH_3CHOH)$ – step *11* – is nothing else than the reverse of a hydride migration to a coordinated aldehyde, see **19**. We can therefore use the results obtained for this migration process by Nakamura and Morokuma [41, 43] and by Versluis [42, 43]. Nakamura and Morokuma looked at the $H_2Ru(CO)_3(CH_2O)$ system using an *ab initio* RHF energy gradient method. Versluis on the other hand considered the $HCo(CO)_3(CH_2O)$ system. His calculations were based on the HFS-LCAO formalism and the energy profile of the reaction was determined through a linear transit procedure. There are two possible pathways for the hydride migration in a $L_nM(H)(CH_2O)$ system, the one leading to the hydroxymethyl complex $L_nM(CH_2OH)$ and the other one leading to the methoxy complex $L_nM(OCH_3)$. The first one is more directly relevant to our problem. It involves (as shown by the calculations [41, 42]) the four centered transition state schematically depicted in **20**.

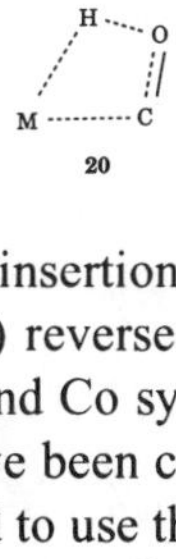

20

Using the values quoted for the insertion reaction [42, 43], one gets for the $L_nM(CH_2OH) \rightarrow L_nM(H)(CH_2O)$ reverse process a barrier of 28.6 kcal/mol and of 3.6 kcal/mol for the Ru and Co systems respectively. It is difficult to compare both studies, which have been carried out on different systems and with quite different methods, and to use their results for a quantitative analysis of the process depicted in **19**. The d^6 six coordinate Ru complex is probably more like the d^8 square planar Pd system. This is best seen from the shapes of the orbitals which are critical for the hydrogen transfer. One finds in all three systems, $HCo(CO)_3(CH_2O)$, $HPd(Cl)(H_2O)(CH_2O)$ and $H_2Ru(CO)_3(CH_2O)$ a formally empty d metal orbital pointing towards the aldehyde and H, see **21–23**. But this empty orbital is of d_{z^2} type

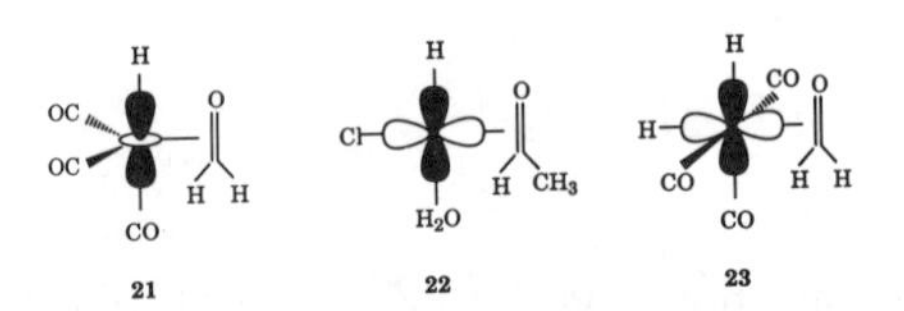

in $HCo(CO)_3(CH_2O)$, see **21**, and is therefore more strongly directed towards H than towards CH_2O. It differs therefore from **22** and **23** which are both of $d_{x^2-y^2}$ type, and both directed towards H and and the aldehyde ligand.

Some useful information can be extracted from the work of Versluis, since it also includes a study, with the same methodology, of the ethylene insertion into the Co-H bond of $HCo(CO)_3(C_2H_4)$. One can therefore, by considering again the reverse process of this ethylene insertion, compare directly the β-H elimination from an alkyl ligand, as in step *9*, to the β-H elimination from an α-hydroxyalkyl ligand, as in step *11*. The two barriers are almost equal, 3.3 kcal/mol for the ethyl complex and 3.6 kcal/mol for the hydroxymethyl complex. There is a difference in the reaction energies however, the β elimination from the ethyl complex being endothermic by about 1.9 kcal/mol whereas the β elimination from the hydroxymethyl is exothermic by 9.5 kcal/mol. It would be now quite interesting to carry out calculations on the actual α and β hydroxyethyl Pd(II) complexes of the steps *9* and *11*.

Finally one may wonder why the acetaldehyde product CH_3CHO, which is obtained at the end of the process, does not insert into the Pd-H bond to give an ethoxy $Pd(Cl)(H_2O)(OCH_2CH_3)$ complex. Indeed in both the Ru and the Co complex discussed above, the methoxy $L_nM(OCH_3)$ system is computed to be *more* stable than the hydroxymethyl $L_nM(CH_2OH)$ system [41–43]. This last feature has been attributed to a more stronger M-O bond as compared to the M-C bond. Here, the palladium complexes may behave differently: we note that in recent and thorough comparative studies of the oxidative addition reaction of H_2O and CH_4 to second row transition metal atom, see the equations *15–16*, Siegbahn *et al.* [44, 45] found that

$$(15) \qquad M + H_2O \rightarrow M(H)(OH) \ (M = Y - Pd)$$

$$(16) \qquad M + CH_4 \rightarrow M(H)(CH_3) \ (M = Y - Pd)$$

the Pd reaction is endothermic by 9.6 kcal/mol and 5.6 kcal/mol for H_2O and CH_4 respectively, and that the Ru reaction is exothermic by 15.8 kcal/mol for H_2O and endothermic by 0.2 kcal/mol for CH_4. Combining these values with the O-H and C-H computed bond dissociation energies (BDE's) in H_2O and

CH_4 [46], one finds that the Pd-(OH) bond is only slightly stronger than the Pd-(CH_3) bond, by 3.1 kcal/mol, whereas the Ru-(OH) bond is much stronger than the Ru-(CH_3) bond, by 23.1 kcal/mol [47]. There is therefore a difference ($\Delta\Delta$BDE) of 20 kcal/mol in favor of the Ru-O bond. A similar comparison can be made from another study of Siegbahn [48] on metal oxide M-O and metal carbene M-CH_2 complexes which yields a difference ($\Delta\Delta$ BDE) of 30.1 kcal/mol again in favor of the Ru-O bond.

The palladium reoxidation steps

As said in the introduction, very little is known about the reoxidation of Pd(0). It seems likely that Pd(0) is oxidized by Cu(II) chloro complexes, perhaps through electron transfer via halide bridges. The function of O_2 is therefore to reoxidize the Cu(I) chloro complexes formed. This Cu(I)$\rightarrow$Cu(II) reoxidation proceeds probably through the formation of an initial oxygen complex and the subsequent formation of radicals such as O_2^-, OH or HO_2, as schematically depicted in equations *17* and *18* respectively [2]. Note that similar mechanisms have been proposed for the

(17) $$CuCl_2^- + O_2 \rightarrow ClCuO_2 + Cl^-$$

(18) $$ClCuO_2 + H_3O^+ \rightarrow CuCl^+ + HO_2 + H_2O$$

reduction of nitric oxide by carbon monoxide catalyzed by an aqueous $PdCl_2$-$CuCl_2$-HCl system [49]. Yet despite their general interest, these reaction have not been studied theoretically. Note that such a study would be hampered at the onset by the fact that the exact nature of the copper chloride complexes in solution is not known with certainty. They can have various stoichiometries, they can be either mononuclear or polynuclear with halide bridges [50]. Calculations have been carried out on $CuCl_2$ and other dihalides [51], CuCl [52], and CuClCO [53]. They show the necessity to take into account the electron correlation effects and also the relativistic effects which are important especially if one wants to describe correctly the weak metal-metal bond interactions in the polymetallic systems. This feature would certainly bring an additional complication in the theoretical study of these reactions.

Conclusion and perspectives

This review has shown that the *ab initio* theoretical studies carried out either on the various steps of the Wacker process or on reactions directly relevant to this problem have been able to rationalize the activation of the coordinated olefin towards the incoming H_2O (or OH^-) nucleophile. In particular, it has been shown that the slipping of the coordinated olefin is central to the activation process and is induced by the incoming nucleophile. The role of the two nucleophilic substitutions which lead from $PdCl_4^{--}$ to the neutral $PdCl_2(H_2O)(C_2H_4)$ intermediate where H_2O and C_2H_4 are *trans* to each other, has been stressed: it allows an easier nucleophilic addition. In contrast, there is a lack of specific theoretical studies devoted to the analysis of the successive β-elimination/insertion steps from the β-hydroxyethyl intermediate. We have been able to use on the other hand the results of related studies to understand why the acetaldehyde product CH_3CHO does not insert into the Pd-H bond to give an ethoxy complex, and the peculiar role of palladium in this context.

Much remains to be done also on the palladium reoxidation steps. Such studies would be of great interest, both experimentally and theoretically.

It is clear finally that all the studies which have been performed so far remain qualitative. A quantitative computational simulation of the steps involved in the Wacker process will require the inclusion of both correlation and relativistic effects. Yet the expected results will be relevant to gas phase reactions only. The next step will be to take into account the solvent effects. It is clear that differential solvation effects should play a great role in the rate determining step *8* and in the substitution steps *4* and *5* as well, although for these last two the solvation effects are probably of outer sphere nature (*vide supra*). One finds here an open field for further theoretical investigations which, in turn, will have implications far beyond the scope of the Wacker process.

A. Dedieu,
Laboratoire de Chimie Quantique,
Université Louis Pasteur, Strasbourg, France

References

1. a) J. Smidt, W. Hafner, R. Jira, J. Sedlmeier, R. Sieber, R. Rüttinger and H. Kojer, Angew. Chem. **71**, 176 (1959); b) J. Smidt, W. Hafner, R. Jira, R. Sieber, J. Sedlmeier and A. Sable, Angew. Chem., **74**, 93 (1962)
2. F.A. Cotton, G. Wilkinson in *A*dvanced Inorganic Chemistry, 5th Edition, p. 1277, J. Wiley, New York (1988)
3. a) J.E. Bäckvall, B. Åkermark and S.O. Ljunggren, J. Chem. Soc. Chem. Comm. 264 (1977); b) J.K. Stille and R. Divakaruni, J. Am. Chem. Soc. **100**, 1304 (1978); c) J.E. Bäckvall, B. Åkermark and S.O. Ljunggren, J. Am. Chem. Soc. **101**, 2411 (1979)

4. O. Eisenstein and R. Hoffmann, J. Am. Chem. Soc. **103**, 4308 (1981)
5. D.M.P. Mingos in Comprehensive Organometallic Chemistry, Chapter 19, G. Wilkinson, F.G.A. Stone and E.W. Abel, Eds., Pergamon, London (1982)
6. A.D. Cameron, V.H. Smith Jr. and M.C. Baird, Int. J. Quantum. Chem. Symp. **20**, 657 (1986); J. Chem. Soc. Dalton 1037 (1988)
7. H. Fujimoto and T. Yamasaki, J. Am. Chem. Soc. **108**, 578 (1986)
8. S. Sakaki, K. Maruta and K. Ohkubo, Inorg. Chem. **26**, 2499 (1987)
9. a) M.J.S. Dewar, Bull. Chem. Soc. Chim. Fr. **18**, C71 (1951); b) J. Chatt and L.A. Duncanson, J. Chem. Soc. 2937 (1953)
10. a) F. Calderazzo and D.B. Dell'Amica, Inorg. Chem. **20**, 1310 (1981); b) R. Uson, J. Forniès, M. Tomas and B. Menjòn, Organomet. **4**, 1912 (1985); c) F. Calderazzo, J. Organomet. Chem. **400**, 303 (1990)
11. L.L. Wright, R.M. Wing and M.F. Rettig, J. Am. Chem. Soc. **104**, 610 (1982)
12. H. Fujimoto, N. Koga and K. Fukui, J. Am. Chem. Soc. **103**, 7452 (1981)
13. The gas phase protonation energies of OH^- and NH_3 are 398 and 212 Kcal/mol. respectively [14]
14. J.E. Del Bene and I. Shavitt, J. Phys. Chem. **94**, 5514 (1994)
15. J.E. Bäckvall, E.E. Björkman, L. Petterson and P. Siegbahn, J. Am. Chem. Soc. **104**, 4369 (1982)
16. a) N. Koga, S. Q. Jin and K. Morokuma, J. Am. Chem. Soc. **110**, 3417 (1988); b) N. Koga and K. Morokuma, in Transition Metal Hydrides, Chapter 6, A. Dedieu, Ed., VCH, New York (1991); c) N. Koga and K. Morokuma, see Chapter 4 of this book
17. This value may be due in part to the Basis Set Superposition Error
18. P.J. Hay, J. Am. Chem. Soc. **103**, 1390 (1981)
19. a) J.N. Dempsey and N.C. Baenziger, J. Am. Chem. Soc. **77**, 4984 (1955); b) H. Suzuki, K. Itoh, Y. Ishii, K. Simon and J.A. Ibers, J. Am. Chem. Soc. **98**, 8484 (1976); c) L. Benchekroun, P. Herpin, M. Julia and L. Saussine, J. Organomet. Chem. **128**, 275, (1977); d) M.F. Rettig, R.W. Wing and G.R. Wiger, J. Am. Chem. Soc. **103**, 2980 (1981); e) A.C. Albeniz, P. Espinet, Y. Jeannin, M. Philoche-Levisalles and B.E. Mann, J. Am. Chem. Soc. **112**, 6594 (1990)
20. R.A. Love, T.F. Koetzle, G.J.B. Williams, L.C. Andrew and R. Bau, Inorg. Chem. **14**, 2653 (1975)
21. S. Sakaki and M. Ieki, Inorg. Chem. **30**, 4218 (1991)
22. P. de Vaal and A. Dedieu, J. Organomet. Chem. **478**, 121 (1994)
23. M. Blomberg, J. Schüle and P.E.M. Siegbahn, J. Am. Chem. Soc. **111**, 6156 (1989)
24. From the reported values of Ref. 23, a thermodynamic cycle can be constructed which allows the determination of the binding energies of C_2H_4 in $PdH_2(C_2H_4)$. The corresponding values are –15 kcal/mol and –24 kcal/mol at the relativistic and non relativistic level respectively. Although the above values indicate that $PdH_2(C_2H_4)$ is non bound with respect to PdH_2 and C_2H_4, they point to a strengthening of the binding by 9 kcal/mol when relativistic corrections are included. For the $Pd(C_2H_4)$ the corresponding values are +7 and –4 kcal/mol. respectively, thus corresponding to a similar increase of 11 kcal/mol
25. More recent and better calculations [26] carried out on the $Pd(C_2H_4)$ system point also to a relativistic effect of 11 kcal/mol. In these calculations a very large basis (e.g. with f polarization functions on the Pd atom) and a thorough evaluation of the correlation effects was included. In particular all valence electrons of C_2H_4 were correlated. This accounts for the much larger C_2H_4 binding energy which is reported, 30.7 kcal/mol and for the quite normal

Pd-C bond length of 2.13 Å, both features being traced entirely to correlation effects. We note here that Sellers [27] in an earlier study also found a large correlation effect (21 kcal/mol from MP2 calculations). The reader might argue that $Pd(C_2H_4)$ is much different from $[PdCl_3(C_2H_4)]^-$ where the correlation effects, according to Sakaki's calculations, are less important. One should nevertheless keep in mind that the orbital interactions (by just sticking to an orbital interaction explanation which is probably valid due to the donation type of bonding encountered here [26]) should not be too different: in both cases the 5s orbital is formally empty for accepting electron density from the $\pi_{C_2H_4}$ orbital, a dσ orbital opposes this forward donation and a doubly occupied dπ orbital is at work for back donation to the $\pi^*_{C_2H_4}$ orbital.

26. M. Blomberg, P.E.M. Siegbahn and M. Svensson, J. Phys. Chem. **96**, 9794 (1992)
27. H. Sellers, J. Comput. Chem. **11**, 754 (1990)
28. A. Olsson and P. Kofod, Inorg. Chem. **31**, 183 (1992)
29. R.J. Cross, Adv. Inorg. Chem. **34**, 219 (1989)
30. B. Åkermark, J. Glaser, L. Öhrström and K. Zettenberg, Organomet. **10**, 733 (1991)
31. A. Dedieu, Topics Phys. Organomet. Chem. **1**, 1 (1985)
32. S.S. Zumdahl and R.S. Drago, J. Am. Chem. Soc. **90**, 6669 (1968)
33. D.R. Amstrong, R. Fortune and P.G. Perkins, Inorg. Chim. Acta , **9**, 9 (1974)
34. A. Rossi and R. Hoffmann, Inorg. Chem. **13**, 365 (1974)
35. J. Burdett, Inorg. Chem. **16**, 3013 (1977)
36. Z. Lin and M.B. Hall, Inorg. Chem. **30**, 646 (1991)
37. D.L. Thorn and R. Hoffmannn, J. Am. Chem. Soc. **100**, 2079 (1979)
38. a) G.M. Whitesides, J.F. Gaasch and E.R. Stedronsky, J. Am. Chem. Soc. **94**, 5258 (1972); b) J.X. McDermott, J.F. White and G.M. Whitesides, J. Am. Chem. Soc. **98**, 6521 (1976); c) R.H. Grubbs, A. Miyashita, M. Liu and P.L. Burk, J. Am. Chem. Soc. **100**, 2418 (1978)
39. N. Koga, S. Obara, K. Kitaura and K. Morokuma, J. Am. Chem. Soc. **107**, 7109 (1985)
40. K. Tatsumi, R. Hoffmann, A. Yamamoto and J.K. Stille, Bull. Chem. Soc. Jap. **54**, 1857 (1981)
41. S. Nakamura and K. Morokuma, Abstracts, 33rd Symposium of Organometallic Chemistry, paper A109, Japan, Tokyo, October 1986
42. L. Versluis, Ph.D. Thesis, The University of Calgary, Canada (1989)
43. A quite detailed account can be found in ref. 16b, pp. 203–206
44. P.E.M. Siegbahn, M.R.A. Blomberg and M. Svensson, J. Phys. Chem. **97**, 2564 (1993)
45. The calculations were carried out with a reasonably large basis set, correlating all valence electrons and including relativistic effects
46. At the above level of calculation, the O-H dissociation energy in water is 115.1 kcal/mol and the C-H binding energy in methane is 108.0 kcal/mol
47. One also makes the additional – and reasonable – assumption that the M-H bond dissociation energy is about the same in M(H)(OH) and in $M(H)(CH_3)$
48. P.E.M. Siegbahn, Chem. Phys. Letters **201**, 15 (1993)
49. a) M. Kubota, K.J. Evans, C.A. Koerntgen and J.C. Marsters Jr., J. Am. Chem. Soc. **100**, 342 (1978); b) M. Kubota, K.J. Evans, C.A. Koerntgen and J.C. Marsters Jr., J. Mol. Cat. **7**, 481 (1980); c) see also K.S. Sun, K.C. Kong and C.H. Cheng, Inorg. Chem. **30**, 1998 (1991)
50. F.A. Cotton and G. Wilkinson in Advanced Inorganic Chemistry, 5th Edition, pp. 757–770, J. Wiley, New York (1988)
51. a) C.D. Gardner, I.H. Hillier and C. Wood, Inorg. Chem. **17**, 168 (1972); b) P. Corrêa de

Mello, M. Hehenberger, S. Larsson and M. Zerner, J. Am. Chem. Soc. **102**, 1278 (1980); c) S. Larsson, B.O. Roos and P.E.M. Siegbahn, Chem. Phys. Letters, **96**, 436, (1983); d) T.K. Ha and N.T. Ngyuen, Z. Naturforsch. **A 39**, 175 (1984); e) S.Y. Shashkin and W.A. Goddard III, J. Phys. Chem. **90**, 255 (1986); f) M.-M. Rohmer, A. Grand and M. Bénard, J. Am. Chem. Soc. **112**, 2875 (1990)

52. a) P. Scharf, S. Brode and R. Ahlrichs, Chem. Phys. Letters, **113**, 447 (1985); b) C. Kölmel and R. Ahlrichs, J. Phys. Chem. **94**, 5536 (1990)
53. H.S. Plitt, M.R.Bär, R. Ahlrichs and H. Schnöckel, Inorg. Chem. **31**, 463 (1992)

R. ZWAANS, J.H. VAN LENTHE AND D.H.W. DEN BOER

A STUDY ON POSSIBLE INTERMEDIATES IN THE EPOXIDATION OF ETHENE CATALYSED BY MANGANESE(III)-CHLORO-PORPHYRIN

1. Introduction

Epoxidation of olefins is used on an industrial scale to convert olefins into useful chemicals. In nature cytochrome P-450 catalyses the epoxidation of olefins by molecular oxygen. The active site of cytochrome P-450 is an iron(III) porphyrin complex. In the laboratory, many model metal porphyrin complexes are able to catalyse olefin epoxidation, though none of them work efficiently with dioxygen as the oxidant. The most active model systems contain either Fe(III) or Mn(III).

The mechanism(s) of metal porphyrin catalysed epoxidation have been the subject of extensive studies, but the problem is by no means solved. The reaction scheme is generally believed to involve transfer of an oxygen atom to the metal center to give some kind of metal-oxo species, which then oxidises the olefin [1]. Neither the structure of the oxo species nor the way in which it oxidises the olefin has been established. Theoretical studies might be able to shed some light on this matter. Therefore, we have undertaken an *ab initio* study of a prototypical manganese(III) porphyrin complex, the oxo species resulting from oxygen transfer to it, and the metal porphyrin-olefin epoxide complex that is assumed to be akin to the final reaction product. In addition we have carried out a number of gradient calculations to investigate possible mechanisms connecting the oxo species with the final product.

Epoxidation is a subject, that has not often been considered using *ab initio* quantum chemical methods. Papers on *ab initio* calculations describing epoxidation catalysed by metals or metal compounds are even less frequently published. The reactions are complicated. Usually there are many side- and consecutive reactions, requiring many compounds and reaction intermediates to be studied. The involvement of exotic reagents makes the use of advanced quantum chemical approaches necessary. It is troublesome to describe the electronic structure of metal atoms involved in the catalysis, even with advanced *ab initio* methods.

Rappé and Goddard [2] report reaction sequences with small chromium, molybdenum and tungsten complexes as catalysts using GVB calculations. Carter and Goddard [3, 4] studied the chemisorption of oxygen on silver

P.W.N.M. van Leeuwen et al. (eds.), Theoretical aspects of homogeneous catalysis, 197–214.

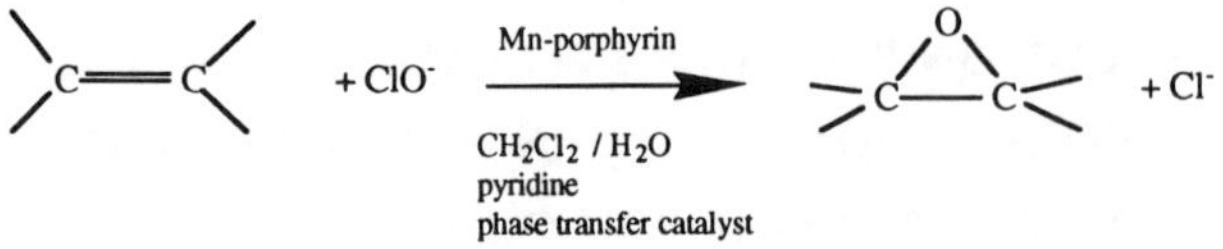

Mn-porphyrin
+ ClO⁻ → + Cl⁻
CH_2Cl_2 / H_2O
pyridine
phase transfer catalyst

Fig. 1. Epoxidation catalysed by manganese-porphyrins.

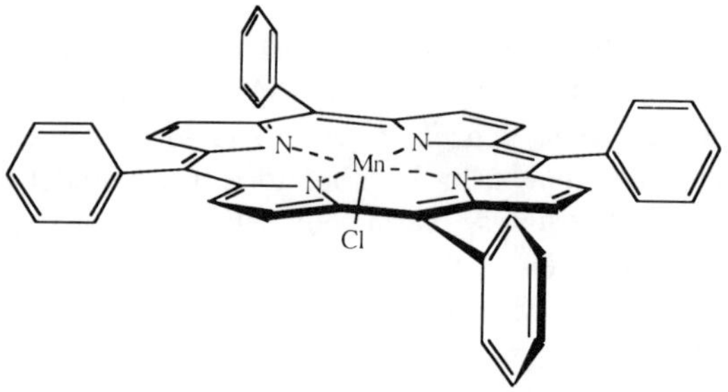

Fig. 2. Manganese(III)-chloro-5,10,15,20-tetra-phenyl-porphyrin.

clusters using the same approach, to deduce a mechanism for the epoxidation reaction. Bach *et al.* [5] used model calculations on LiOOH to deduce the importance of low lying empty d-orbitals in a metal catalyst. Hofmann and Clark [6] suggest that Li^+ diminishes the barriers of epoxide formation from a single oxygen atom and an ethene molecule. They suppose that the transition of oxygen towards the double bond occurs via an asymmetrical pathway.

There is a wealth of experimental evidence on the ability of manganese-porphyrins to catalyse the epoxidation of double carbon-carbon bonds [7–9]. With hypochlorite as oxygen donor the overall reaction can be represented as in Figure *1*.

Porphyrins are aromatic molecules; the π-electrons of the carbon and nitrogen atoms form a large, delocalised π-system. The structure of a typical catalyst, manganese(III)-chloro-5,10,15,20-tetra-phenyl-porphyrin, is shown in Figure *2*.

^{18}O labelling studies [8] have shown, that one of the intermediates in the catalytic cycle contains one oxygen atom per active species. Trapping experiments with DPPH (1,1-diphenyl-2-(2,4,6-trinitro-phenyl)hydrazine), show that this species is two oxidising equivalents above the Mn(III) resting state [10]. The actual oxidation site could not clearly be pointed out: whether it was on the metal atom or (somewhere) on the porphyrin moiety. The extreme instability of this oxo species prevents an adequate characterisation with experimental methods.

An interesting review on *ab initio* calculations on metal-porphyrins was written by Dedieu *et al.* [11]. It describes calculations of quite high quality, especially for the time (1982). Minimal and some extended basis-set Hartree-Fock calculations are considered for Fe-, Mn-, Co- and Ti-porphyrin complexes even including some dioxygen complexes. The groundstate electronic configurations are found to be high spin. Since no gradient techniques were practical at that time, full geometry optimisations were not feasible and much information from experimental structures had to be incorporated in the calculations. Only a small number of specific parameters could be optimised, like the O-O bondlength in the dioxygen complex, and they used idealised geometries exhibiting as much symmetry as possible.

Our *ab initio* investigation started with the determination of the geometry of the oxo-intermediate. At first a structure for manganese(III)-oxo-chloro-porphyrin with C_{4v} symmetry was suggested [12]. Within this constraint the geometry was optimised. The oxygen atom and the chlorine ligand atom were both attached to the manganese atom, each on opposite sites of the porphyrin plane.

Later studies showed that this structure does not represent a minimum conformation. We found a species containing an oxygen atom bridging between the manganese atom and one of the nitrogen atoms [13] (reducing the symmetry to C_s). This structure agrees with X-ray data [14], electronic spectra [15], and *ab initio* [16] as well as semi-empirical calculations [17, 18]. Finally the spin-states of the complexes were systematically considered and ethene was added to the complex [19].

We present Hartree-Fock calculations on three intermediates in the catalysed epoxidation reaction of hypochlorite with ethene. We give the calculated geometry and charge distributions of manganese(III)-chloro-porphyrin and manganese(III)-oxo-chloro-porphyrin. The addition of an ethene molecule to the oxo-particle results in a manganese(III)-epoxide-chloro-porphyrin complex. On the basis of these calculations we give some suggestions about the catalytic cycle.

2. Computational approach

The calculations presented in this chapter were performed using the Restricted Hartree-Fock formalism incorporated in the quantum chemical program system GAMESS-UK [20–22]. For each atom a minimal STO3G [23–25] basis set was used. In order to describe the metal atom on the same level of

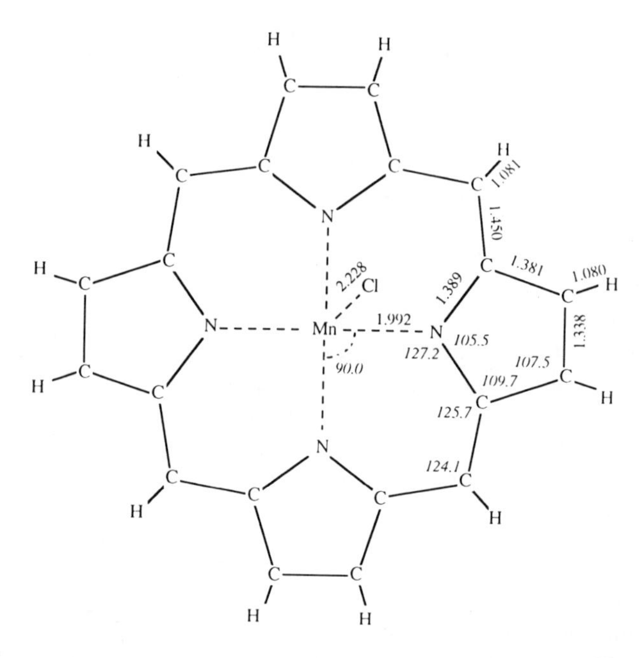

Fig. 3. Top view of the geometry of manganese(III)-chloro-porphyrin with distances and angles (*italics*).

accuracy as the other atoms in the complex, it turned out to be essential to augment the STO3G basis set on the manganese atom with an extra set of d-Gaussians ($\zeta = 0.26$) [12]. The charge distributions are obtained using a Distributed Multipole Analysis [26, 27].

The model catalyst studied here contains a porphyrin moiety without substituents. A chlorine atom is used as a ligand. This leaves us 45 atoms in the largest complex (manganese(III)-epoxide-chloro-porphyrin) with a total of 226 electrons divided over 185 basis functions. In all calculations the porphyrin moieties were assumed to be planar.

All geometric parameters are given in Ångstrøms and degrees. The charges are given in atomic units (an electron has charge −1) and the energies are given in hartrees: 1 hartree = 2625.5 kJoule/mol = 627.52 kcal/mol.

The calculations were performed on the group's HP750 and HP/Apollo DN10000, the Silicon Graphics Challenge of the faculty of chemistry and the Cray-YMP at SARA (Amsterdam).

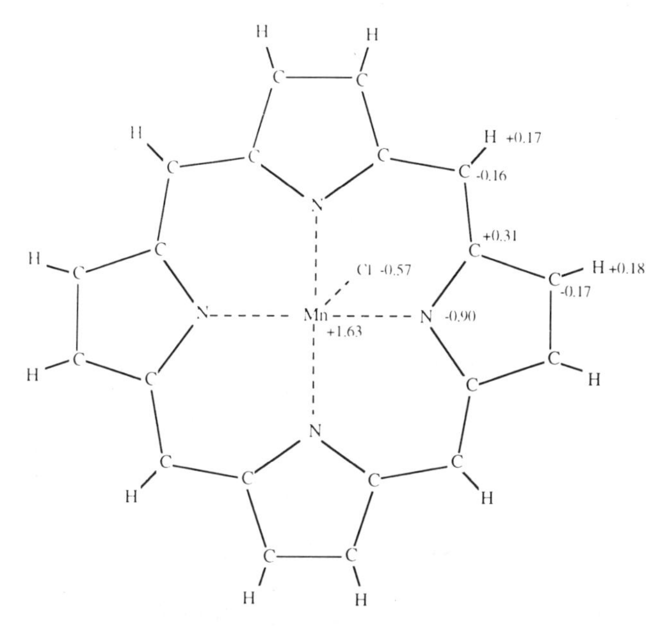

Fig. 4. Charge distribution of manganese(III)-chloro-porphyrin; charges in electrons. Charges not shown are determined by the C_{4v}-symmetry.

3. Results

3.1. MANGANESE(III)-CHLORO-PORPHYRIN

The starting point in the calculations is the "bare catalyst". Figure *3* shows the optimised geometry of manganese(III)-chloro-porphyrin (C_{4v} symmetry). The ground state configuration is a quintet spin state, containing four singly occupied orbitals (cf. Figure *12*).

The charge distribution of manganese-chloro-porphyrin is given in Figure *4*. The charges on the atoms are remarkably high. Charges of this size are possible because the charge distribution is stabilised by rings of positive (the manganese atom, the inner carbon atoms and the hydrogens) and negative charge (the nitrogen atoms and the outer ring of carbon atoms). Still, the formal valency of the manganese atom may not be interpreted as a charge of +3.

In the small basis we use for these calculations, the separation of the charges is actually underestimated. Improving the basis will increase the polarisation of the charges. For example, when the manganese, nitrogen and chlorine

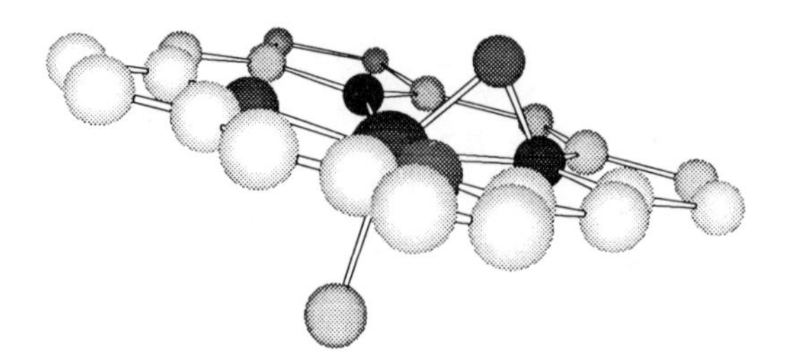

Fig. 5. The geometry of manganese(III)-oxo-chloro-porphyrin.

atom are described with a Split Valence (SV) 3–21G basis [28, 29], the charge on manganese already increases from +1.63 to +2.19 [19].

3.2. MANGANESE (III)-OXO-CHLORO-PORPHYRIN

Like manganese-chloro-porphyrin the ground state of manganese-oxo-chloro-porphyrin has quintet spin multiplicity (cf. Figure *12*). This is in contrast to our earlier work [12, 13], where a closed shell configuration was assumed. We even found an excited state of the complex with triplet spin multiplicity and a lower energy than the lowest singlet state.

The removal of all symmetry constraints from manganese(III)-oxo-chloro-porphyrin leads to a geometry with the oxygen atom bridging between the manganese atom and one of the nitrogen atoms of the porphyrin moiety. The compound possesses nearly C_s symmetry. The porphyrin ring is still constrained to be planar, which may have had considerable influence on the results. The geometry of the complex does not differ very much from our earlier result [13]. Figure *5* shows an impression of the complex.

The oxygen atom is not located straight above the manganese atom, but it has shifted towards one of the nitrogen atoms to form a nitrogen-oxygen bond.

Figure *6* shows the bond lengths and angles in manganese(III)-oxo-chloro-porphyrin. The distortions within the porphyrin ring due to the influence of the oxygen atom are not very large, but the chlorine atom moves away from the oxygen atom to the opposite quadrant of the porphyrin system.

The charge distribution of manganese(III)-chloro-porphyrin can still be recognised in the charge distribution of manganese(III)-oxo-chloro-porphyrin, but now it is perturbed by the presence of the oxygen atom. The charge effects

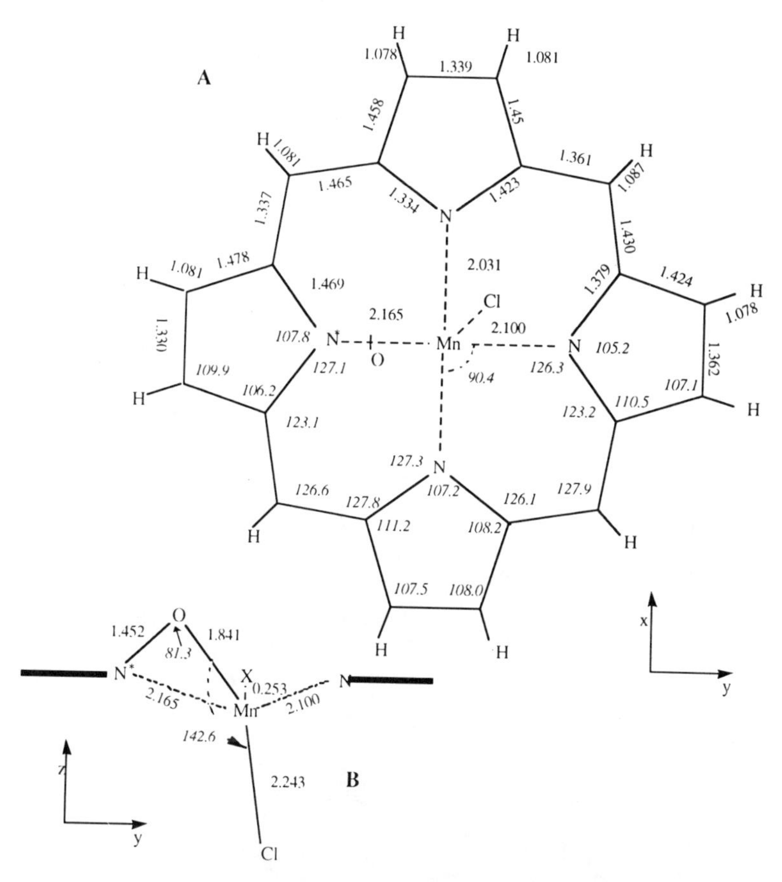

Fig. 6. Schematic representation of the calculated geometry of manganese(III)-oxo-chloro-porphyrin with distances and angles (*italics*). A: top view, B: side view.

resulting from the addition of oxygen to manganese(III)-chloro-porphyrin are mainly local effects. The formation of the nitrogen-oxygen-metal bridge involves an oxidation of the nitrogen involved in the bridge; the porphyrin ring moiety takes care of the surplus.

So the oxidation of manganese(III)-chloro-porphyrin by the oxygen atom takes place on the bridging nitrogen atom of the porphyrin ring moiety. Its charge decreases from –0.90 to –0.45. This accounts for the major part of the

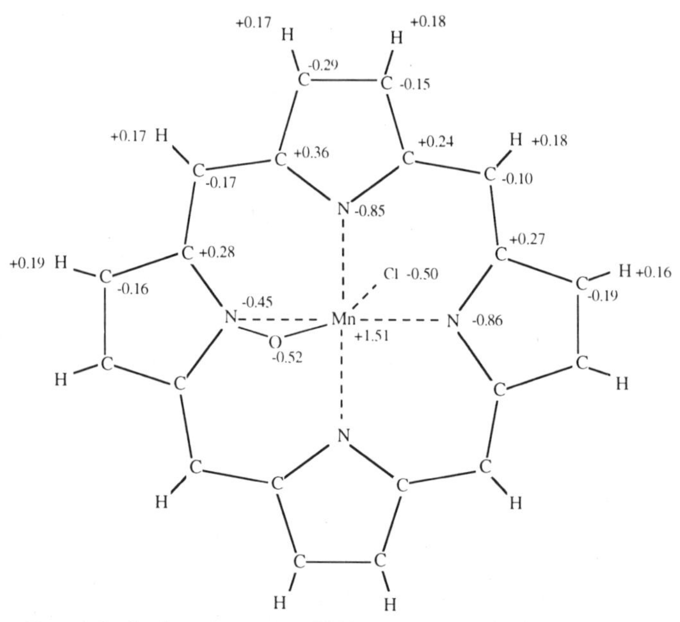

Fig. 7. Charge distribution of manganese(III)-oxo-chloro-porphyrin. Charges not shown are fixed by the C_s-symmetry.

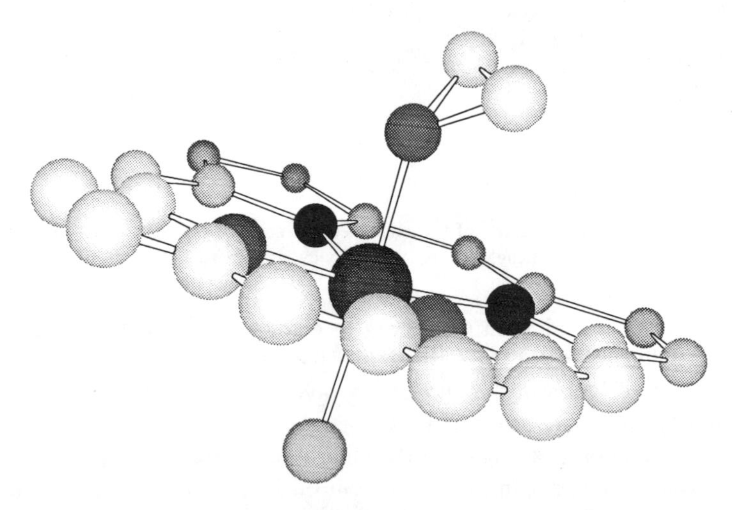

Fig. 8. The geometry of manganese(III)-epoxide-chloro-porphyrin.

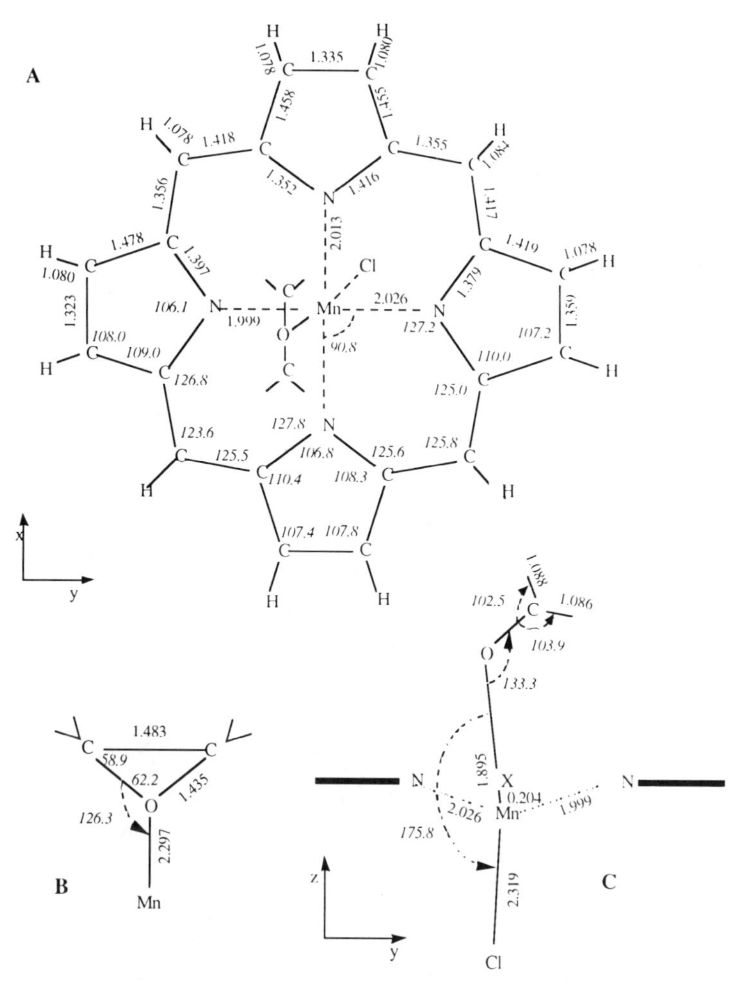

Fig. 9. Geometry of manganese(III)-epoxide-chloro-porphyrin with distances and angles (*italics*). A: top view, B,C: side views.

reduction of the oxygen atom, that becomes –0.52 charged. The manganese atom is only used to coordinate the oxygen atom; it does not donate negative charge to the oxygen atom. Its charge even slightly decreases from +1.63 to +1.51. The binding between the metal atom and the porphyrin is weakened

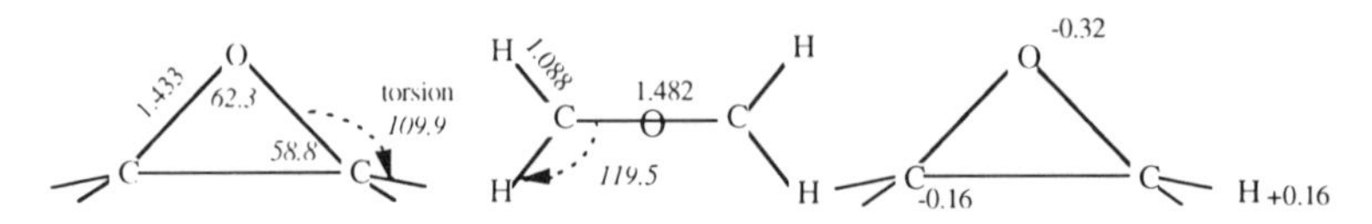

Fig. 10. Calculated geometry and charge distribution (right) of a single epoxide molecule.

by the oxidation as shown by the longer metal-nitrogen bonds. The double Mn=O metal-oxygen bond, which is a common representation for the complex, is not found in our calculations.

Oxidation of the metal porphyrin complex at the ligand rather than at the metal would have been expected on the basis of frontier-orbital arguments, since the HOMO of the complex is a nearly pure ligand orbital [30, 31]. (cf. Figure *12*).

3.3. MANGANESE(III)-EPOXIDE-CHLORO-PORPHYRIN

The addition of an ethene molecule to the intermediate manganese(III)-oxo-chloro-porphyrin molecule results in a complex of manganese(III)-chloro-porphyrin with an epoxide ligand. This structure is shown in Figure *8*. This is not necessarily the only or optimal conformation for the complex. The epoxide molecule is situated perpendicular to a manganese-nitrogen bond.

The oxygen atom is situated almost straight "above" the geometrical centre of the porphyrin moiety with the metal-oxygen "bond" approximately perpendicular to the porphyrin plane. The bridging bond between nitrogen and oxygen is broken. The chlorine atom is situated "underneath" the oxygen atom with an almost linear chlorine-manganese-oxygen axis. The geometrical parameters are given in Figure *9*.

The differences with the geometry of manganese(III)-chloro-porphyrin are very small: they are within hundreds of Ångstrøms for the bond lengths or within a few degrees for the angles.

The only significant difference between the geometry of the coordinated epoxide and the free oxirane molecule is in the torsion angles the hydrogen atoms make with the carbon-oxygen-carbon plane (104.3° and 101.9° vs. 109.9° in a single epoxide molecule (Figure *10*)).

When the epoxide/manganese(III)-chloro-porphyrin ligand complex is formed, the charge distribution of the porphyrin is almost equal to that of

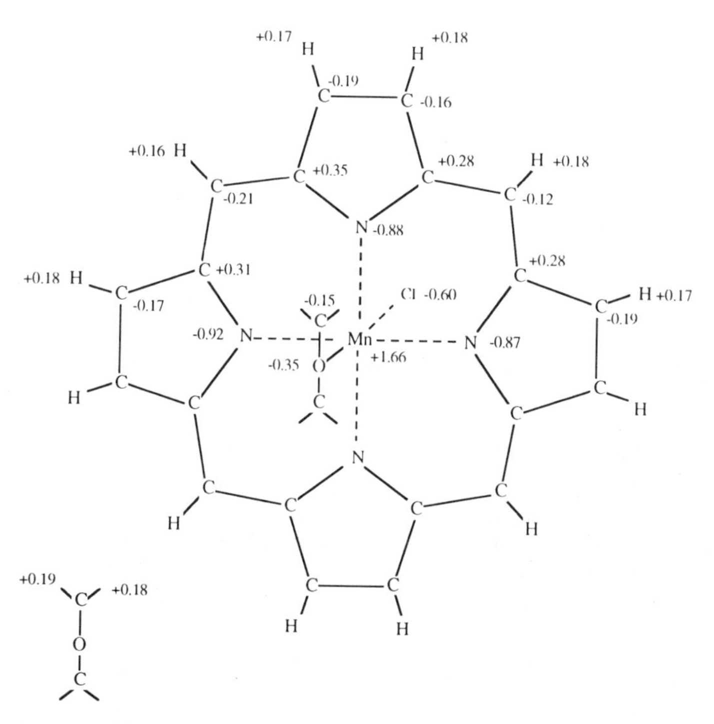

Fig. 11. Charge distribution of manganese(III)-epoxide-chloro-porphyrin. The charge distribution obeys C_s symmetry.

manganese(III)-chloro-porphyrin; the porphyrin has returned to its original oxidation state. The charge distribution of manganese(III)-epoxide-chloro-porphyrin is given in Figure *11*.

The breaking of the oxygen-nitrogen bond is illustrated by the charge of the nitrogen atom: with a charge of –0.87 it is almost back to its state in manganese(III)-chloro-porphyrin. The charges on manganese (changed by +0.03) and chlorine (changed by –0.03) also hardly differ.

The epoxide ligand is almost neutral; its total charge is +0.09. So in this state of the complex there is no transfer of electrons from the porphyrin ring to the oxygen atom. The charge distribution within the epoxide ligand resembles the charge distribution of a single epoxide molecule (cf. Figure *10*).

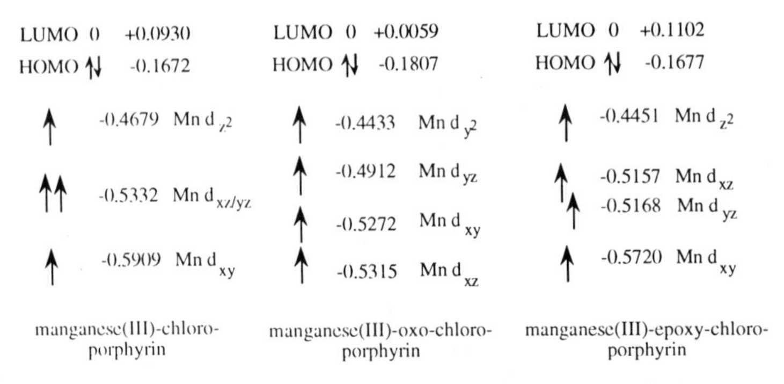

Fig. 12. Electronic configuration of the manganese(III)-complexes. See Figure *9* for the direction of the axes.

All this suggests an epoxide ligand, that is rather loosely complexed with the positively charged manganese atom in a manganese-chloro-porphyrin. The complex, shown in Figure *8*, is the final result of the geometry optimisation. No conclusion about reaction pathways may be drawn from these calculations. However, in exploratory calculations we have not found stable complexes of ethene with the manganese-oxo-chloro-porphyrin.

3.4. SPIN STATES OF THE MANGANESE(III)-CHLORO COMPLEXES

All complexes have quintet spin-symmetry. Configuration Interaction calculations on manganese(III)-chloro-porphyrin using a SV 3–21G basis set [28, 29] for the chlorine, metal and nitrogen atoms, confirm this [19]. The orbital energies and character of the singly occupied orbitals, as deduced from their largest AO-coefficients, is shown in Figure *12*. The orbitals in the oxo species are not as clearly identifiable as in the other complexes, but are somewhat rotated. The orbital denoted here as d_{y^2} is actually mainly a mixture of .87 d_{y^2}, an orbital like a d_{z^2} in the y direction, and .29 d_{yz} orbital. The d_{y^2} orbital contains .52 d_{z^2} in addition to .72 d_{yz}.

The HOMO and the LUMO are orbitals belonging to the porphyrin ring π-system, mainly consisting of carbon and nitrogen p_π orbitals.

The main characters of the singly occupied orbitals and the sequence of their energies does not vary between the different complexes. The major Atomic Orbital (AO) coefficients that describe the singly occupied orbitals are manganese d-coefficients, but there is some contamination from other manganese orbitals.

In the oxo complex one can characterise the singly occupied orbitals less clearly, because the presence of the oxygen atom disrupts the symmetry of the system. The four unpaired electrons remain localised on the manganese atom. So the singly occupied orbitals are not involved in the binding of the oxygen atom to the complex. Instead, they provide an antibonding force, thereby loosening the metal-oxygen bond. Therefore, this bond effectively becomes only half a bond which may facilitate the epoxidation reaction.

The addition of a closed shell ethene molecule to the oxo-porphyrin compound does not change the total spin multiplicity. The character of the singly occupied orbitals in manganese(III)-epoxide-chloro-porphyrin is easier to establish than in manganese(III)-oxo-chloro-porphyrin, because here the manganese atom is almost square planar surrounded. Thus the occupation pattern shows much similarity with the orbital occupation of manganese(III)-chloro-porphyrin. The Mn d_{xz} and Mn d_{yz} orbitals are nearly degenerate and the manganese d-coefficients are by far the largest coefficients present.

Other spin states were examined. The geometry of the complexes does not differ very much between different spin multiplicities. This can be explained from the metal-character of the singly occupied orbitals. A local spin flip on the manganese atom does not influence the atoms of the porphyrin moiety.

If one considers the system as a whole, the "Aufbau" principle is obviously not adhered to. If the manganese atom is considered separately from the rest of the complex, there is Aufbau within each separate part. The manganese d-orbitals have little interaction with the porphyrin ring. Considering these orbitals in the three complexes, a picture emerges of a Mn^{3+} ion with a Cl^- counter-ion. This system is then stabilised by the surrounding porphyrin^{2-} ion. In the Mn^{3+} ion the singly occupied d-orbitals are basically degenerate and their energies are affected by the surrounding charge distribution.

4. Considerations on the epoxidation reaction

4.1. THE ATTACK OF ETHENE ONTO MANGANESE(III)-OXO-CHLORO-PORPHYRIN

The attack of an ethene molecule onto manganese(III)-oxo-chloro-porphyrin was investigated for several positions of the ethene molecule relative to the manganese-oxygen-nitrogen bridge. These calculations were performed, using a closed-shell configuration for the complex.

By calculating the energy gradients of the nuclear coordinates of the oxygen atom and the carbon atoms of the ethene molecule, the forces working on the ethene and oxygen moieties were estimated and thus the direction of their possible drift. Positioning of the ethene molecule near the manganese atom did not result in a possible motion of oxygen and ethene towards each other. When the ethene molecule was situated perpendicular to the manganese-oxygen-nitrogen bridge gradients pointing towards epoxidation were obtained. Optimisation of this geometry resulted in the epoxide complex presented in Section 3.4. These results suggest that a metal centred attack of the ethene molecule is not likely to happen.

4.2. IS THE TRANSITION STATE SYMMETRIC?

Both the reaction intermediates manganese(III)-oxo-chloro-porphyrin and manganese(III)-epoxide-chloro-porphyrin have C_s or near C_s symmetry. This suggests a symmetrical attack of the ethene molecule on the manganese-oxygen-nitrogen bridge and a symmetrical particle at the height of the reaction barrier. Experimental evidence [8] shows, however, that *cis-trans* isomerisation of symmetrically substituted alkenes occurs during the reaction: a *cis* substituted alkene results both in *cis* and in *trans* substituted epoxides

The *cis-trans* isomerisation requires activation energy to overcome a reaction barrier. It does not occur when the substituents are small (e.g. *cis*-2-octene) or when voluminous side chains are present at the porphyrin moiety. Isomerisation from *cis*-alkenes to *trans*-epoxides happens more often than from *trans*-alkenes to *cis*-epoxides, even when the *cis/trans* equilibrium for the substitution of epoxide is taken into account. Collman *et al.* [32] observe pseudo-stereospecificity for *trans*-2-methylstyrene, *trans*-2-octene and *trans*-4-octene, because oxidation of these *trans*-isomers already leads to the more stable epoxide.

In this work we consider a single Mn(III) → [Mn(III) + O] → Mn(III) + epoxide catalytic cycle. There are however indications that the Meunier [32] manganese porphyrin system actually has two competing reaction pathways, one of which involves a "radical" Mn(IV) intermediate. The isomerisation commonly observed during epoxidation with this system has been attributed to the Mn(IV) pathway, cf. [33, 34].

4.3. REACTION ENERGIES

We now consider the reaction of ethene with a hypochlorite ion to form epoxide and a Cl^- ion, catalysed by manganese-porphyrin as is depicted in Figure *1*. The total energies of the compounds involved in the epoxidation are given in Table *I*.

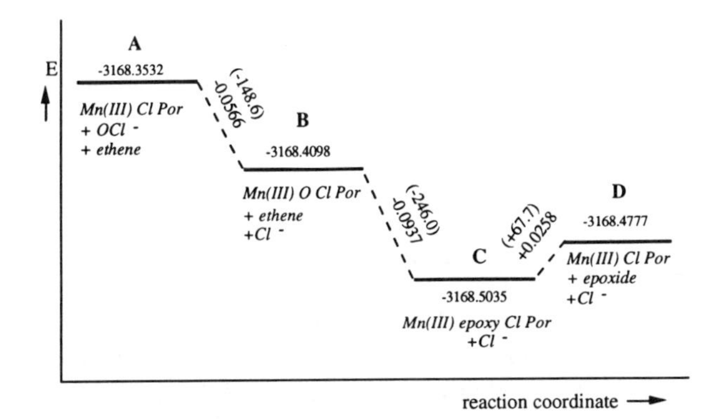

Fig. 13. Reaction energies of the complexes and of the reaction steps corresponding to the particles evaluated in this chapter (in atomic units and kJ/mole (between brackets)).

TABLE I
Total energies and spin ground states of the reactants

Compound	Spin state	Total energy (hartree)
Hypochlorite anion (1-)	closed shell singlet	–528.210490
Chlorine anion (1-)	closed shell singlet	–454.480422
Ethene	closed shell singlet	–77.073954
Epoxide	closed shell singlet	–150.928503
Mn(III) Cl Por	quintet	–2563.068748
Mn(III) O Cl Por	quintet	–2636.855458
Mn(III) epoxide Cl Por	quintet	–2714.023104

From these data we calculate the reaction energies given in Figure *13*. The spin multiplicity throughout the reaction sequence is constant (quintet). The reactants, hypochlorite (OCl^-), ethene and epoxide have singlet spin multiplicity. No estimate is made of energy barriers between the steps nor is an absence of barriers implied.

Two processes occur between {**A**} and {**B**}. The oxygen-chlorine bond in hypochlorite is broken (+0.1731 hartree = 454.5 kJ/mol) and two new bonds are formed: the oxygen-manganese and oxygen-nitrogen bonds in manganese(III)-oxo-chloro-porphyrin (this releases 0.2297 hartree = 603.1 kJ/mol).

The oxo-particle {**B**} is a metastable intermediate. During the formation of the epoxide molecule (between {**B**} and {**C**}) energy is released. The epoxide particle {**C**} is one of the minima of the reaction sequence, so the epoxide ligand has to be moved away by some outside influence (e.g. by thermal energy or a collision with a molecule from the solvent). The porphyrin is then ready to start a new reaction cycle after removal of the epoxide molecule at {**D**}.

5. Conclusions

The oxidation of manganese(III)-chloro-porphyrin by a hypochlorite occurs in both senses of the word on the porphyrin moiety. The oxygen atom becomes attached to the porphyrin. It assumes a bridging position between the manganese atom and one of the nitrogen atoms. The resulting oxo-complex has almost C_s symmetry. It has a quintet spin ground state at SCF level.

The reduction of the oxygen atom is performed by the nearest nitrogen atom. The transfer of electrons to the oxygen atom is stabilised by the porphyrin moiety. This includes the manganese atom: its charge decreases slightly because it is situated next to the nitrogen atom.

An ethene molecule will attack the oxo-particle described above perpendicular to the manganese-oxygen-nitrogen bridge. The attack results in a manganese(III)-epoxide-chloro-porphyrin complex. This complex also has a quintet spin ground state. The singly occupied orbitals do not change their character.

When the complex has been formed, the porphyrin moiety has returned in its original oxidation state. The epoxide ligand is held by the manganese atom, that has nearly the same charge as in the oxo-complex and in manganese(III)-chloro-porphyrin.

The structure of the manganese-porphyrin, that is suggested by the results of these calculations, is that of a metal atom solvated by a porphyrin ring. The porphyrin ring is rigid, not allowing a close binding with the manganese atom. The metal-atom's structure is that of a Mn^{3+} atom, where only the energies of the singly occupied d orbitals are influenced by the surroundings.

This work was sponsored by the Stichting Nationale Computerfaciliteiten (National Computing Facilities Foundation, NCF) for the use of supercomputer facilities, with financial support from the Nederlandse Organisatie voor Wetenschappelijk Onderzoek (Netherlands Organization for Scientific Research, NWO).

We thank the staff of the SARA computer center, especially Dr. L.C.H. van Corler, for support. We thank Prof. L.W. Jenneskens, Prof. P.W.N.M van Leeuwen and Dr. P.H.M. Budzelaar for helpful discussions.

This work forms part of the Ph.D. research of R. Zwaans, which has carried out with financial support from Kominlilyke Shell Laboratorium, Amsterdam (KSLA).

R. Zwaans, J.H. van Leuthe and D.H.W. den Boen,
Theoretical Chemistry Group, Debye Institute,
Utrecht University, Padualaan 14, 3584 CH Utrecht, The Netherlands

References

1. W.A. Nugent and J.M. Mayer, Metal-Ligand multiple bonds, pp. 248–252 Wiley Interscience, New York, (1988)
2. A.K. Rappe and W.A. Goddard III, J. Am. Chem. Soc. **102**, 5114 (1980)
3. E.A. Carter and W.A. Goddard III, Surf. Sci. **209**, 243 (1989)
4. E.A. Carter and W.A. Goddard III, J. Catal. **112**, 80 (1988)
5. R.D. Bach, G.J. Wolber and B.A. Coddens, J. Am. Chem. Soc. **106**, 6098 (1984)
6. H. Hofmann and T. Clark, Angew. Chem. **102**, 697 (1990)
7. K.A. Jørgensen, Chem. Rev. **3**, 431 (1989)
8. A.W. van der Made, *O*n the Epoxidation of Alkenes with Hypochlorite Catalysed by Manganese(III)tetraarylporphyrins, Ph.D. Thesis, Utrecht (1988)
9. D. Ostovic and T.C. Bruice, Acc. Chem. Res. **25**, 314 (1992)
10. A.W. van der Made, P.M.F.C. Groot, R.J.M. Nolte and W. Drenth, Recl. Trav. Chim. Pays-Bas **108**, 73 (1989)
11. A. Dedieu, M.-M. Rohmer and A. Veillard, Adv. Quantum Chem. **16**, 43 (1982)
12. D.H.W. den Boer, J.H. van Lenthe and A.W. van der Made, Recl. Trav. Chim. Pays-Bas **107**, 256 (1988)
13. D.H.W. den Boer, A.W. van der Made, R. Zwaans and J.H. van Lenthe, Recl. Trav. Chim. Pays-Bas **109**, 123 (1990)
14. A.L. Balch, Y.-W. Chan and M.M. Olmstead, J. Am. Chem. Soc. **107**, 6510 (1981)
15. B. Chevrier, R. Weiss, M. Lange, J.C. Chottard and D. Mansuy, J. Am. Chem. Soc. **103**, 2899 (1981)
16. A. Strich and A. Veillard, Nouv. J. Chim. **7**, 347 (1983)
17. K.A. Jørgensen, J. Am. Chem. Soc. **109**, 698 (1987)
18. K. Tatsumi and R. Hoffmann, Inorg. Chem. **20**, 3771 (1981)
19. R. Zwaans, *A*b Initio Organic Chemistry, Calculations on Transition Metal Complexes, Ph.D. Thesis, Utrecht (1993)
20. M. Dupuis, D. Spangler and J.J. Wendoloski, *N*RCC Software Catalog Programme, Vol 1. Program No QG01 (GAMESS) (1980)
21. M.F. Guest, R.J. Harrison, J.H. van Lenthe and L.C.H. van Corler, Theoretica Chimica Acta (Berl.) **71**, 117 (1987)
22. M.F. Guest, P. Fantucci, R.J. Harrison, J. Kendrick, J.H. van Lenthe, K. Schoeffel and P.

Sherwood, GAMESS-UK, CFS Ltd, Daresbury (1993)
23. W.J. Hehre, R.F. Stewart and J.A. Pople, J. Chem. Phys. **51**, 2657 (1969)
24. W.J. Hehre, R. Ditchfield, R.F. Stewart and J.A. Pople, J. Chem. Phys. **52**, 2769 (1970)
25. W.J. Pietro and W.J. Hehre, J. Comput. Chem. **4**, 241 (1983)
26. A.J. Stone, Chem. Phys. Lett. **83**, 233 (1981)
27. A.J. Stone and M. Alderton, Mol. Phys. **56**, 1047 (1985)
28. J.S. Binkley, J.A. Pople and W.J. Hehre, J. Am. Chem. Soc. **102**, 939 (1980)
29. K.D. Dobbs and W.J. Hehre, J. Comput. Chem. **8**, 861 (1987)
30. K. Fukui (1970) Theory of Orientation and Stereoselection, in Fortschr. Chem. Forsch., Vol 15, Springer Verlag
31. I. Fleming, *F*rontier Orbitals and Organic Chemical Reactions. Wiley-Interscience, New York (1976)
32. J.P. Collman, J.I. Brauman, B. Meunier, T. Hayashi and T. Kodadek, J. Am. Chem. Soc. **107**, 2000 (1985)
33. J.T. Groves and M.K. Stern, J. Am. Chem. Soc. **109**, 3812 (1987)
34. R.D. Arasasingham, G.-X. He and T.C. Bruice, J. Am. Chem. Soc. **115**, 7985 (1993)

INDEX

Catalysis by Metal Complexes

Series Editors:

R. Ugo, *University of Milan, Milan, Italy*

B. R. James, *University of British Columbia, Vancouver, Canada*

1.* F. J. McQuillin: *Homogeneous Hydrogenation in Organic Chemistry*. 1976 ISBN 90-277-0646-8
2. P. M. Henry: *Palladium Catalyzed Oxidation of Hydrocarbons*. 1980 ISBN 90-277-0986-6
3. R. A. Sheldon: *Chemicals from Synthesis Gas*. Catalytic Reactions of CO and H_2. 1983 ISBN 90-277-1489-4
4. W. Keim (ed.): *Catalysis in C_1 Chemistry*. 1983 ISBN 90-277-1527-0
5. A. E. Shilov: *Activation of Saturated Hydrocarbons by Transition Metal Complexes*. 1984 ISBN 90-277-1628-5
6. F. R. Hartley: *Supported Metal Complexes*. A New Generation of Catalysts. 1985 ISBN 90-277-1855-5
7. Y. Iwasawa (ed.): *Tailored Metal Catalysts*. 1986 ISBN 90-277-1866-0
8. R. S. Dickson: *Homogeneous Catalysis with Compounds of Rhodium and Iridium*. 1985 ISBN 90-277-1880-6
9. G. Strukul (ed.): *Catalytic Oxidations with Hydrogen Peroxide as Oxidant*. 1993 ISBN 0-7923-1771-8
10. A. Mortreux and F. Petit (eds.): *Industrial Applications of Homogeneous Catalysis*. 1988 ISBN 90-277-2520-9
11. N. Farrell: *Transition Metal Complexes as Drugs and Chemotherapeutic Agents*. 1989 ISBN 90-277-2828-3
12. A. F. Noels, M. Graziani and A. J. Hubert (eds.): *Metal Promoted Selectivity in Organic Synthesis*. 1991 ISBN 0-7923-1184-1
13. L. I. Simándi: *Catalytic Activation of Dioxygen by Metal Complexes*. 1992 ISBN 0-7923-1896-X
14. K. Kalyanasundaram and M. Grätzel (eds.), *Photosensitization and Photocatalysis Using Inorganic and Organometallic Compounds*. 1993 ISBN 0-7923-2261-4

Catalysis by Metal Complexes

15. P. A. Chaloner, M. A. Esteruelas, F. Joó and L. A. Oro: *Homogeneous Hydrogenation.* 1994 ISBN 0-7923-2474-9
16. G. Braca (ed.): *Oxygenates by Homologation or CO Hydrogenation with Metal Complexes.* 1994 ISBN 0-7923-2628-8
17. F. Montanari and L. Casella (eds.): *Metalloporphyrins Catalyzed Oxidations.* 1994 ISBN 0-7923-2657-1
18. P.W.N.M. van Leeuwen, K. Morokuma and J.H. van Lenthe (eds.): *Theoretical Aspects of Homogeneous Catalysis.* Applications of *Ab Initio* Molecular Orbital Theory. 1995 ISBN 0-7923-3107-9

KLUWER ACADEMIC PUBLISHERS – DORDRECHT / BOSTON / LONDON

**Volume 1 is previously published under the Series Title: Homogeneous Catalysis in Organic and Inorganic Chemistry.*